CONSTRUCTION DICTIONARY POCKET EDITION

**An Abridged Edition Of
"Building News
Illustrated Construction Dictionary"**

Published By

BNi Building News

Division Of BNI Publications, Inc.

BNI®Building News

BNI PUBLICATIONS, INC.
1-800-873-6397

LOS ANGELES	ANAHEIM
10801 National Blvd.	1612 S. Clementine St.
Los Angeles, CA 90064	Anaheim CA 92802

BOSTON	WASHINGTON,D.C.
629 Highland Ave.	502 Maple Ave. West
Needham, MA 02194	Vienna, VA 22180

ISBN 1-55701-181-8

A

Abandonment. The failure of both parties to a contract to abide by its terms.

Above Ground Tank. A large above ground vessel used for the storage of liquids.

Abrasion. Wearing away by friction.

Abrasion Resistance. Ability of a surface to resist being worn away by rubbing and friction.

Abrasive. A substance used for wearing, grinding, or rubbing away by friction.

Abrasive Surface. A surface that has been roughened for safety or for warning.

Abrasive Surface Tile. Floor tile that has been roughened to be slip-resistant.

ABS. Acrylonitrile-butadiene-styrene, hard plastic used because of its resistance to impact, heat and chemicals.

ABS Pipe. A plastic pipe sold in 10 and 20 foot lengths in different diameters for plumbing stacks and drains. Used primarily for drain lines.

Absorption. The relationship of the weight of the water absorbed by a material specimen subjected to prescribed immersion procedure, to the weight of the dry specimen, expressed in percent.

AC. Initials for alternating current. Alternating current is delivered to a building or structure from the local utility company. Alternating current flows alternately in one direction, then the opposite, completing 60 cycles every second.

Acceleration. Rate of change of velocity.

Accelerator. A substance which, when added to concrete, mortar, or grout, increases the rate of hydration of the hydraulic cement, shortens the time of setting, or increases the rate of hardening of strength development, or both. Materials used to speed up the setting of mortar or concrete.

Acceptance. Manifestation that a party assents or agrees to a contract.

Access Door. A door or panel creating a means of access for the inspection or repair of concealed equipment.

Access Floor. A raised floor platform with removable panels to provide access to the area below.

Access Stair. A stair system to provide specific access to roofs, mechanical equipment rooms or as a means of exit in an emergency.

Accessories (Tile Accessories). Ceramic or non-ceramic articles, affixed to or inserted in tile work, as exemplified by towel bars, paper, soap and tumbler holders, grab bars and the like.

Accessory. An object or device aiding or contributing in a secondary way.

Accessory, Concrete. An implement or device used in the formwork, pouring, spreading, finishing, etc. of concrete surfaces.

Accessory, Reinforcing. The items used to assemble or facilitate in the installation of masonry or concrete reinforcing.

Accord and Satisfaction. Conduct of a debtor that indicates agreement to an amount of money owed by the debtor to a creditor.

Accordion Folding Door. A folding, hinged or creased door with rollers which run along a track.

Accordion Partition. A folded, creased, or hinged interior dividing wall.

Acid. A chemical substance usually corrosive to common metals (iron, aluminum, zinc) and which, in water solution, imparts an acid, sour or tart taste.

Acid Etch. The use of acid to cut lines into metal or glass. The use of acid to remove the surface of concrete.

Acid Rain. Sulfur dioxide emissions combining with water in the atmosphere and falling to the earth.

Acid-Proof Counter. A horizontal work surface resistant to acid spills.

Acidity. A general term applying to substances on the acid side of neutral - principally the degree of acidity.

ACM. Asbestos-Containing Material. Any material containing more than one percent asbestos.

Acoustical Treatment. The act or process of applying acoustical materials to walls and ceilings.

Acrylic. A general class of resinous polymers derived from esters, amides or other acrylic aid derivatives. A transparent plastic material used in sheet form for window glass and skylights.

Act of God. An unexpected event, not within the control of either party, that makes the performance of a contract impossible, unreasonable, or illegal.

Active Pressure. The pressure exerted by retained earth; such as the earth retained by a retaining wall.

AD Plywood. A designation or gradation of plywood. The "A" and the "D" designate quality of surface layers.

Adapter. A device for connecting two different parts.

Addenda. A revision in the contract document made prior to the execution of the owner-contractor contract.

Additive. A term frequently used as a synonym for addition or admixture.

Adhesive. A substance that dries or cures and binds two surfaces together. A substance capable of holding materials together by surface attachment.

Adjustable Bar Hanger. A metal hanger that can be made to fit the varying distances between floor and ceiling joists or rafters to securely hold electrical outlet boxes and devices.

Adjustable Shelf. A shelf that can be adjusted to different heights.

Adjustable Shelf Hardware. Metal items to provide for the support of shelves in multiple positions.

Adjustable Shelf Standard. Metal items to support shelves usually in the form of strips attached to vertical surfaces.

Admixture. A material other than water, aggregates, and hydraulic cement used as an ingredient of concrete or mortar, and added immediately before or during its mixing. A chemical additive used to alter the normal properties of concrete.

Adobe. Unburned or unfired brick, dried in the sun.

ADR. Alternative Dispute Resolution. Resolution of a dispute without litigation.

Affidavit. A written statement that is made under oath.

Agent. One who is authorized to act on behalf of a principal.

Aggregate. Inert particles such as sand, gravel, crushed stone, or expanded materials, in a concrete or plaster mixture. Granular material, such as sand, gravel, crushed stone, and iron blast-furnace slag, used with a cementing medium to form a hydraulic-cement, concrete or mortar.

Air Compressor. A tool which takes air and forces it at a high pressure into a storage tank. The air is released through a regulator and a hose to power small tools.

Air Conditioning. Equipment with an apparatus for controlling the humidity and temperature of air.

Air Distribution. To force air to desired locations in a building or facility.

Air Eliminator. A mechanical device that expels excess air.

Air Entraining Agent. A substance added to concrete, mortar or cement that produces air bubbles during mixing, making it easier to work with and increasing its resistance to frost and freezing.

Air Handling. Single or variable-speed fans pushing air over hot or cold coils, through dampers and ducts to heat or cool a building or structure.

Air Handling System. A system to heat, cool, humidity, dehumidify, filter and transport air. It consists of an air handling unit, fresh air (F.A.) and exhaust air (E.A.) damper at the building exterior, ductwork, supply air (S.A.), diffusers or registers, and return air (R.A.) grills in the conditioned space.

Air Plenum. Any space usually used to convey return air in a building or structure.

Air Powered Hoist. A hoist that is operated by compressed air.

Air-Slack. A condition where soft-body clay, after absorbing moisture and being exposed to the atmosphere, will spall a piece of clay and/or glaze.

Air Tool. Attachments using compressed air to saw, spray-paint, sand, drill or nail, etc.

Air Vent. An opening in a building or structure for the passage of air.

Airborne Sound. Sound originating in a space. Also, airborne sound can be created from the radiation of structure-borne sound into the air.

All Risk Policy. A property insurance policy that insures against all risks of loss that are not specifically excluded.

Alkali. A chemical substance which effectively neutralizes acid material so as to form neutral salts. A base. The opposite of acid. Examples are ammonia and caustic soda.

Allowance. In the contract documents, a sum noted by the architect to be included by the contract sum, for a specific item. For example, a stipulated sum for carpeting or hardware which will be selected at a later date.

Alloy. A substance composed of two or more metals, or of a metal and a nonmetallic constituent.

Alternating Current (AC). An electrical current that has the voltage difference reversing its direction at a fixed frequency.

Alternative Dispute Resolution. Resolution of a dispute without litigation.

Aluminum. A bluish silver-white malleable metallic element with good electrical and thermal conductivity, high reflectivity and resistance to oxidation.

Aluminum Faced Tile. A thin, rectangular unit that has a sheet of aluminum bonded to it.

Aluminum Flashing. Material in sheets constructed of aluminum used to cover open roof joints in exterior construction to make waterproof.

Aluminum Plate. Flat aluminum sheet material.

Aluminum Siding Panel. Rectangular sheets of aluminum, used to cover the exterior surface of the outside of a building or structure.

Aluminum Storefront. A facade of a building or structure which is constructed of a system of aluminum tubing and glass.

Aluminum Window. A unit placed in an opening in a wall for light and ventilation with its framing constructed from aluminum.

Ambient Sound. The quiet-state noise level in a room or space, which is a composite of sounds from many external sources, both near and far, over which one individual has no control.

Ambiguous. Having more than one meaning, e.g. In a contract.

Ammeter. An instrument for measuring electric current in amperes.

Amperes (Amp). A unit of measure of electrical flow in a conductor.

Amplifier. A device to obtain amplification of voltage, current or power.

Amplitude. Maximum deviation from mean or center line of a wave.

Analysis. Separation into constituent parts. In engineering, the investigative determination of the detailed aspects of a particular phenomenon.

Anchor. Irons or metals of special form and shapes used to fasten together and secure timbers or masonry.

Anchor Bolt. A bolt embedded in concrete for the purpose of fastening a building frame to a concrete or masonry foundation.

Anchor, Rafter. A bolt or fastening device which attaches the rafters used to support a roof to the walls or rafter plate.

Anchor Slot. A groove in an object into which a fastener or connector is inserted to attach objects together.

Anchorage. An attachment for resistance to movement. The movement can be a result of overturning, sliding or uplift. The most common anchorage for these movements are tie-downs (hold-downs) for overturning and uplift, and anchor bolts for sliding.

Anchoring Cement. Grout used in sleeves to anchor tubing in place.

Angle. A structural section of steel which resembles an "L" in cross section.

Angle Valve. A valve in which the shut-off in the pipe openings are set at right angles to each other.

Annealed. Cooled under controlled conditions to minimize internal stresses.

Annunciator Panel. A panel mounted on a surface which indicates by lights which circuits have been activated.

Anodized. A metal that has been subjected to electrolytic action in order to coat with a protective or decorative film.

Antenna. A metallic device used for radiating or receiving radio waves.

Anti-Siphon. A device to prevent the removal of fluid caused by suction produced by fluid flow.

Anticipatory Breach. A positive statement by a party to a contract that the party will not perform the terms of the contract.

Anticlastic. Saddle-shaped, or having curvature in two opposing directions.

Antimicrobial. Agent that kills microbial growth.

Anti-Siphon. A device to prevent the removal of fluid caused by suction produced by fluid flow.

Application for Payment. A written document forwarded by the contractor requesting payment for work completed.

Apron. The lower trim member under the sill of the interior casing of a window. An upward or downward vertical extension of a sink or lavatory. A paved area immediately adjacent to a building, structure or facility.

Arbitration. A proceeding for resolution of disputes in which a neutral person, after hearing evidence presented by both sides, makes a final and binding decision that resolves the dispute. A hearing used to resolve disputes.

Arc Welding. A process of joining two pieces of metal by melting them together at their interface with a continuous electric spark and adding a controlled additional amount of molten metal from a metallic electrode.

Arch. A curved structure in which compression is the essential cause of internal stresses. A structural member with an upward-curved centerline axis and a relatively small cross section, supported so that the distance between its ends cannot change. Both horizontal and vertical components of reactive forces are therefore found at both ends. Forces active normal to the arch center line are carried in compression by virtue of its curvature.

Arch Culvert. A curved shaped drain under a roadway, canal or embankment.

Architectural Equipment. The implements, apparatus, or equipment used in the construction and initial outfitting of a building.

Architectural Fee. The amount of money charged for the preparation of plans with specifications for the construction of a building or facility.

Architectural Woodwork. Finish work using wood or composites for ornamental designs or casework.

Armor Plate. A kick-plate made of metal installed on the bottom of a door to protect it from denting and scratching.

Arrester, Lightning. A device connected to an electrical system to protect from lightning and/or voltage surges.

Asbestos. A fibrous mineral that was used for its noncombustible, nonconducting or chemically resistant properties.

As Builts. A set of drawings prepared by the general contractor, which includes any revisions in the working drawings and specifications during construction, indicating how the project was actually constructed.

Ashlar Veneer. An ornamental or protective facing of masonry composed of squared stones.

Asphalt. A residue in petroleum or coal-tar refining that is used for pavements and as a waterproofing cement.

Asphalt Shingle. Saturated roofing felt either in large rolls or cut into composition shingles, impregnated with aggregate particles applied to a roof surfaces.

Assignment. Transfer of the rights and duties under a contract from one party to another.

Astragal Weatherstripping. Fabric, rubber or plastic strips attached to the molding that is attached to one of a pair doors or casement windows to cover up the joint between the two stiles.

Attenuation. The reduction of the energy or intensity of sound.

Attenuation Blanket. Material utilized to help in the reduction of the energy or intensity of sound.

Attic Insulation. Treated shreds of cellulose material that are blown into attic spaces or fiberglass rolls that are rolled out between ceiling joists to aid in weatherproofing a building or facility.

Auger. An instrument or device used for boring or forcing through materials or soil.

Autoclave. A pressure vessel in which an environment of steam at high pressure may be produced; used in the curing of concrete products and in the testing of hydraulic cement.

Auxiliary Switch. A standby device for switching.

Award. A written decision signed by an arbitrator that resolves a dispute that has been submitted to arbitration by the parties to the dispute.

Awning Window. A window hinged at the top.

Axial Load. Force directly coincident with the primary axis of a structural member such as a beam.

Axis. A straight line of reference. In three dimensions one usually refers to three axes, x, y, and z.

Axis of Symmetry. A line dividing an area into two similar but opposite handed figures.

Axis, Neutral. Centroidal axis, transverse to longitudinal axis of a structural member, which is neither stretched nor shortened by bending of the member.

B

Back Splash. A protective panel installed behind a counter, sink or lavatory.

Back-Up Block. Block which is used for the inner load bearing, structural portion of a masonry wall.

Back-Up Brick. A load bearing or structural portion of a masonry wall constructed of brick against which a veneer is attached.

Backcharge. An offsetting charge against a bill, often asserted by an owner against a prime contractor or a by prime contractor against a subcontractor based on supposedly defective construction work.

Backfill. Earth or earthen material used to fill the excavation around a foundation; the act of filling around a foundation.

Backflow. The unintentional flow of water or other substances into the distribution pipes of a water drainage system from a source other than the intended source.

Backhoe. An excavating machine with a bucket rigidly attached to a hinged stick on a boom that is drawn toward the machine in operation.

Backhoe/Loader. An excavation machine combining a bucket on a hinged stick on a boom on one end, and a bucket or scoop at the other.

Backup Bar. A small rectangular strip of steel applied beneath a joint to provide a solid base for beginning a weld between two steel structural members.

Balanced Cuts. Cuts of tile at the perimeter of an area that will not take full tiles. The cuts on opposite sides of such an area shall be the same size. Also the same sized cuts on each side of a miter.

Ball Clay. A secondary clay, commonly characterized by the presence of organic matter, high plasticity, high dry strength, long verification range, and a light color when fired.

Ball Valve. A valve in which a ball regulates the opening by its rise and fall due to fluid pressure, a spring, or its own weight.

Ballast. A device used with a fluorescent and high intensity lamp, to provide the necessary circuit condition for starting and operation. Any material used as non-structural fill or dead weight.

Ballast, Roof. Crushed rock or gravel which is spread on a roof surface to form its final surface.

Balloon Frame. A wooden building frame composed of closely spaced members (studs) which are continuous from the sill to the top plate of the roof line.

Baluster. Vertical members that extend from a stair, or floor, to a handrail. A small pillar or column used to support a rail.

Balustrade. A series of balusters connected by a rail, generally used for porches, balconies and the like.

Band. A low, flat molding.

Band Joist. A wooden joist perpendicular to the direction of the joists in a floor, closing off the floor platform at the outside face of the building.

Band Saw. A power sawing machine using a saw in the form of an endless steel belt running over pulleys.

Bank Run Gravel. Excavated material that is generally 1/4 inch minimum to 6 inches maximum.

Bankruptcy. A legal proceeding by which a debtor may avoid legal and financial obligations.

Bar, Reinforcing. A manufactured usually deformed steel bar, used in concrete and masonry construction to provide additional strength.

Barbed Wire. Wire that is twisted with barbs or sharp points.

Bare Solid Wire. Un-insulated single wire used as an electric conductor.

Bare Stranded Wire. Un-insulated group of fine wires used as a single electric conductor.

Barge. A floating platform or vessel from which construction activities may be performed. Often used in rivers to install bridge piers and also used extensively in waterfront construction.

Barge Board. The installation of ornamental boards to conceal roof timbers projecting over gables.

Barrel Shell. A scalloped roof structure of reinforced concrete that spans in one direction as a barrel vault and in the other as a folded plate.

Barrel Vault. A segment of a cylinder that spans as an arch. Used as a structural technique to support a ceiling or roof by having all of the components act in compression as an "arched" ceiling. Used extensively in ancient buildings and into the 19th century because no structural steel or timber is needed.

Barricade. An obstruction to prevent passage or to prevent access.

Barrier, Vapor. A type of plastic sheeting that both eliminates drafts and keeps moisture from damaging a building or structure.

Base. The bottom of a column; the finish of a room at the junction of the walls and floor. One or more rows of tile installed above the floor.

Base Bid. A stipulated construction sum based on the contract documents.

Base Cabinet. Case, box, or piece of furniture which sets on floor with sets of drawers or shelves with doors, primarily used for storage.

Baseboard Heater. Heating strips that are installed at the juncture of the wall and floor and may be either recessed or surface-mounted; generally along the outside walls of rooms.

Baseplate. A steel plate inserted between a column and the foundation, used to level the column and to spread the load of the column to a larger area of the foundation.

Basis for Acceptance. The method of determining whether a lot of material is acceptable under given or accepted specifications.

Batch Mixer. A machine which mixes batches of concrete or mortar in contrast to a continuous mixer.

Batch Plant. An operating installation of equipment including batchers and mixers as required for batching or for batching and mixing concrete materials; also called mixing plant when equipment is included.

Bath. A tub used for bathing. The room that contains the tub.

Batt Insulation. Loosely matted plant or glass fibers with one or both sides faced with kraft paper or aluminum foil available in specifically sized sections.

Batten Siding. Vertical siding which has narrow strips of metal or wood covering the joints.

Batter Board. A temporary framework used to assist in locating the corners when laying a foundation. Also used to maintain proper elevations of structures, excavations and trenches in any kind of below ground construction.

Batter Pile. Pile driven at an angle to brace a structure against lateral or horizontal thrust.

Bay. A rectangular area of a building defined by four adjacent columns; a portion of a building that projects from a facade.

Bead. A narrow line of weld metal or sealant; a strip of metal or wood used to hold a sheet of glass in place; a narrow, convex molding profile; a metal edge or corner accessory for plaster.

Beam. A straight structural member that acts primarily to resist transverse loads. A structural element which sustains transverse loading and develops internal forces of bending and shear in resisting the loads. An inclusive term for joists, girders, rafters, and purlins.

Beam, Grade. An end-supported horizontal load-bearing foundation member that supports an exterior wall.

Beam, Reinforcing. A horizontal member installed to strengthen and support the load of a structure.

Bearing Wall. A wall which supports any vertical loads in addition to its own weight.

Bedding. A filling of mortar, putty, or other substance in order to secure a firm bearing.

Bedrock. A solid layer or stratum of rock beneath ground level.

Bell and Spigot. Cast iron pipe joints formed with sections that have a wide opening (bell) at one end and a narrow end (spigot) at the other. They are then fitted by caulking with oakum and lead.

Bellows. An instrument or machine that draws in air through a valve or orifice by expansion and contraction and expels it through a tube.

Belt Course. A horizontal board across or around a building, usually made of a flat member and a molding.

Bench Mark. Permanent reference point or mark.

Bench Saw. A power saw held securely on a stationary bench.

Bend, Soil. A piece of short, curved pipe, like an elbow, used to connect two straight links of conduit in a sewage system.

Bending Moment. The sum of moments for all forces that occur above the neutral axis. The moment that causes a beam or other structural member to bend.

Bentonite. A clay composed principally of minerals of the montmorillonoid group, characterized by high absorption and very large volume change with wetting or drying, commonly swelling to several times its dry volume when saturated with liquid.

Bevel. An end or edge cut at an angle other than a right angle.

Bevel Board (Pitch Board). A board used in framing roof or stairway to lay out bevels.

Bevel Siding. Wood siding that tapers in cross section.

Beveled Concrete. An angle in concrete or inclination of any line in concrete or concrete surface that joins another.

Bibb, Hose. A water spigot or faucet with its nozzle threaded or coupling attached to accept a hose.

Bid Bond. A bond, secured by the bidder, which guarantees that the bidder selected by the owner will accept the project, or the owner will have the project for the bid price as noted in the accepted bid.

Bid. An offer to perform; an offer to enter into a contract usually for a stipulated sum of money.

Bid Shopping. A general contractor contacts subcontractors in an attempt to receive a lower subcontractor price after having been awarded the contract for the project.

Bi-Fold Door. A door with two leaves, hinged together to close on itself. One edge of each leaf is hinged at the jamb and the other edge is connected and guided by an overhead track.

Bin Method. A method of computing cooling energy use requirements for commercial and industrial building with unusual operating needs and for residences utilizing passive heating/cooling design with high mass thermal storage.

Birch Veneer. Thin sheets of strong fine-grained hardwood used in furniture, flooring, etc.

Birdscreen. Wire screening attached to louvers, ventilators and openings in a building or structure to prevent birds and small animals from entering.

Bitumen. A tar based mixture, such as asphalt or coal tar.

Bituminous. Resembling, containing, or impregnated with various mixtures of hydrocarbons (like tar) together with their nonmetallic derivatives.

Bituminous Sidewalk. A walkway constructed with an impregnated mixture of hydrocarbons together with aggregate such as sand or stone. Commonly called "blacktop".

Blanket Insulation. Thermal insulating material made of fibrous glass or mineral wool, sometimes offered with paper or foil surfacing, formed in batts or rolls.

Bleeding. The flow of mixing water within, or its emergence from newly placed concrete or mortar; caused by the settlement of the solid materials within the mass; also called water gain.

Blind Nailing. Attaching boards to framing or sheathing with nails driven through the edge of each piece so as to be concealed by the board.

Block, Concrete. A hollow concrete masonry unit constructed a composite material consisting of sand, coarse aggregate, cement and water.

Block, Glass. A hollow masonry unit made of glass.

Block, Granite. A masonry unit consisting of a very hard natural igneous rock used for its firmness and endurance.

Block, Grout. Mortar mixes used in block walls to fill voids and joints.

Block, Splash. A small masonry block placed in the ground beneath a downspout to receive roof drainage and prevent standing water or soil erosion.

Block, Terminal. A decorative element forming the end of a block structure.

Block Vent. An opening serving as an outlet or inlet for air in a block structure.

Blocking. Pieces of wood inserted tightly between joists, studs, or rafters in a building frame to stabilize the structure, inhibit the passage of fire, provide a nailing surface for finish materials, or retain insulation.

Blown-In Insulation. Insulation made of cellulose (paper treated with a fire retardant) that is blown into an attic, crawl space or walls by a blowing machine.

Bluestone. A sandstone of a dark-greenish to bluish-gray color that splits into thin slabs, commonly used to pave surfaces for pedestrian traffic.

Board. Lumber less than 2 inches thick.

Board Foot. A unit of lumber volume, a rectangular solid nominally 12" x 12" x 1". The equivalent of a board 1 foot square and 1 inch thick.

Board Siding. A type of lumber installed on the exterior walls of a building or structure to act as the finish sheathing.

Board Sub-Flooring. A wooden member that is installed on floor joists to which the finished floor is fastened.

Boarding In. The process of nailing boards on the outside studding of a house.

Boiler. A closed vessel in which heat is produced by combustion or electricity and transferred to water. A heating system in which water is used as the distribution medium.

Bollard. Short steel post (usually filled with concrete) set to prevent vehicular access to or to protect property from damage by vehicular encroachment. Steel or cast iron post to which ships are tied.

Bolster. A long wire type "chair" used to support reinforcing bars in a concrete slab while the concrete is being placed.

Bolt. A fastener consisting of a cylindrical metal body with a head at one end and a helical thread at the other, intended to be inserted through holes in adjoining pieces of material and closed with a threaded nut.

Bolted Steel. Steel members bolted with a metallic pin or rod having a head at one end and threads on the other for attaching the nut.

Bond. The adherence of one material to another. Effective bonds must be achieved between the mortar and scratch coat, between the tile and mortar, and between the adhesive and backing.

Bond Beam. A horizontal grouted element within masonry in which reinforcement is embedded.

Bond Breaker. A material used to prevent adhesion of newly placed concrete to other surfaces.

Bond Coat. A material used between the back of the tile and the prepared surface. Suitable bond coats include pure portland cement. Dry-Set portland cement mortar, latex-type portland cement mortar, organic adhesive, and the like.

Bond, Roof. A legal guarantee that a roof installed is in accordance with specifications and will be repaired or replaced if it fails in a certain period of time due to normal weathering.

Bonding Agent. A substance applied to a suitable substrate to create a bond between it and a succeeding layer as between a subsurface and a terrazzo topping or a succeeding plaster application.

Bookkeeper. A person who records the accounts or transactions of a business.

Booth, Spray. An area in a building or structure used for spray painting. Blocked off by walls to prevent dust and dirt from landing on work surface.

Bored Lock. A door lock manufactured for installation in a circular hole in a door.

Boring. To make holes in wood or metal to aid in the insertion of bolts, nails or other fasteners. To drill into the ground to bring up samples of earth for testing.

Borrow. Excavated material that has been taken from one area to be used as fill at another location.

Bottle Cooler. A container used for cooling or maintaining the coolness of bottled liquids.

Bottom Bars. The reinforcing bars that lie close to the bottom of a reinforced concrete beam or slab.

Bottom Beam. The lowest horizontal member supporting a building or structure.

Bottom Pivot Hinge. A flexible pair of plates joined by a pin to allow swinging of a door or gate installed at the bottom.

Bottom Plate. A flat horizontal member, also called a mudsill, that supports the vertical wall studs and posts. Horizontal wood lumber member at bottom of wall studs which ties them together and supports studs, and which rests on the sill or joists.

Boundaries. A separating line that indicates or fixes a limit or extent.

Bowl, Toilet. The oval part of a toilet which receives the waste and fills with water after flushing the toilet tank.

Box Beam. A beam of metal, concrete or plywood which, in cross section, resembles a closed rectangular box.

Box Culvert. A concrete drainage structure rectangular shaped, reinforced and cast in place or made of precast sections.

Box, Distribution. A box which contains the circuit breakers, connects to the service wires and delivers current to the various outlets throughout a building or structure.

Box, Floor. An electrical, metal outlet box providing outlets from conduits in or under a floor.

Box, Gang. Electrical boxes constructed of metal or hard plastic, manufactured with knockout holes to pull wire through to connect outlets or switches.

Braced Frame. A truss system or its equivalent which resists lateral forces.

Braced Wall Line. A series of braced wall panels in a single story.

Bracing. Diagonal members, either temporary or permanent, installed to stabilize a structure against lateral loads. Structural member used to prevent buckling or rotation of wood studs.

Bracket. A projecting support for a shelf or other structure.

Branch Breaker. A switch which stops the flow of current by opening the circuit automatically when more electricity flows through the circuit than the circuit is capable of carrying. Resetting may be either automatic or manual.

Brass Fitting. Threaded pipe connector constructed of brass, used to join two pieces of pipe together.

Brazed Connection. Parts that are hardened and connected by soldering with an alloy.

Breach of Contract. A material failure to perform an act required by contract.

Break Joints. To arrange joints so that they do not come directly under or over the joints of adjoining pieces, as in shingling, siding, etc.

Breaker, Circuit. A switch which stops the flow of current by opening the circuit automatically when more electricity flows through the circuit than the circuit is capable of carrying. Resetting may be either automatic or manual.

Breaker, Main. A switch in a main electrical service panel where the two hot service wires attach.

Brick. A solid masonry unit having the shape of a rectangular prism. Usually made from clay, shale, fire clay, or a mixture of these.

Brick Anchor. Fasteners that are designed to attach and secure a brick veneer to a concrete or brick wall.

Brick, Paver. Brick units that are used in foot traffic areas. They are four inches wide, eight inches long, and 1-5/8 to 2-1/4 inches thick.

Bridge Crane. A hoisting device spanning two overhead rails. The hoisting device moves laterally along the bridge with the bridge moving longitudinally along the rails.

Bridge Deck. The slab or other structure forming the travel surface of a bridge.

Bridging. Pieces fitted in pairs from the bottom of one floor joist to the top of adjacent joists, and crossed to distribute the floor load; sometimes pieces of width equal to the joists and fitted neatly between them. Diagonal or longitudinal members used to keep horizontal members properly spaced, in lateral position, vertically plumb, and to distribute load.

Bronze. An alloy of copper and tin and sometimes other elements.

Bronze Valve. Any of numerous mechanical devices by which the flow of liquid, gas (as air) or loose material in bulk is controlled, manufactured from an alloy of copper and tin and sometimes other elements.

Broom Finish. A finish applied to an uncured concrete surface, to provide skid or slip resistance, made by dragging a broom across the freshly placed concrete surface.

Brown Coat. The second of three coats of a plaster or stucco application.

Brownstone. A brownish or reddish sandstone.

Brush Cutting. The act of removing unwanted plants to clear an area.

Brushed Surface. A sandy texture obtained by brushing the surface of freshly placed or slightly hardened concrete with a stiff brush for architectural effect or, in pavements, to increase skid resistance.

BTU. British thermal unit; measurement of the heat energy required to raise one pound of water one degree Fahrenheit.

Bucket Trap. A mechanical steam trap operating on buoyancy that prevents the passage of steam through the mechanical system it protects.

Buckle. To bend under compression. With very thin members, the bucking may be elastic, and the member will spring back if the load is removed. If the load is continued or if the buckling occurs with the stresses above the yield point, the member will fail by collapsing completely.

Building Envelope. Elements of the building, including all external building materials, windows, and walls, that enclose the internal space.

Building, Metal. A building or structure constructed of a structural steel frame covered by metal roof and wall panels. Commonly prefabricated in a factory and assembled at the site.

Building Official. The official charged with administration and enforcement of the applicable building code, or his duly authorized representative.

Building Paper. Water repellent paper used to assist in shedding incidental moisture what may penetrate exterior finishes of exterior wall construction. Thick paper, used to insulate a building before the siding or cladding is installed. Occasionally used in flooring between double floors.

Built-Up Member. A single structural component made from several pieces fastened together.

Built-Up Roof. A roof covering made of continuous rolls or sheets of saturated or coated felt. They are then cemented together with bitumen and may have a final coating of gravel or slag.

Built-Up Steel Lintel. Lintel fabricated of two or more pieces of structural steel secured together to act as one member.

Built-Up Timber. A timber made of several pieces fastened together, an forming one of larger dimension.

Bulb Tee. Rolled steel in the form of a "T" with a formed bulb on the edge of the web.

Bulk Excavation. The digging out of large amounts of dirt and debris.

Bulkhead Formwork. The temporary formwork that blocks fresh concrete from a section of forms or closes the end of a form at a construction joint.

Bulking. Increase in the bulk volume of a quantity of sand in a moist condition over the volume of the same quantity dry.

Bulking Curve. Graph of change in volume of a quantity of sand due to change in moisture content.

Bull Float. A tool comprising a large, flat, rectangular piece of wood, aluminum, or magnesium usually 8 in. (20 cm) wide and 42 to 60 in. (100 to 150 cm) long, and a handle 4 to 16 ft. (1 to 5 m) in length used to smooth unformed surfaces of freshly placed concrete.

Bulldozer. A tractor driven machine with a horizontal blade for clearing land, road building, or similar work.

Bullnose. A trim tile with a convex radius on one edge. This tile is used for finishing the top of a wainscot or for turning an outside corner.

Bumper, Dock. Thick rubber units placed under loading dock openings to absorb the shock and prevent damage when trucks back in for loading or unloading.

Bumper, Door. Rubber tip devices mounted on walls or baseboards that prevent door knobs from marring walls.

Bundled Bars. A group of not more than four parallel reinforcing bars in contact with each other, usually tied together.

Buoyant Uplift. The force of water or liquefied soil that tends to raise a building foundation out of the ground.

Burlap. A coarse fabric of jute, hemp, or less commonly, flax, for use as a water-retaining covering in curing concrete surfaces; also called Hessian.

Burlap Rub. A finish obtained by rubbing burlap to remove surface irregularities from concrete.

Bus Duct. A prefabricated unit containing one or more electric conductors, often a metal bar, that serves as a common connection for two or more circuits.

Bush Hammer. A hammer used to dress concrete or stone with its serrated face and may pyramidal points. A hammer that has a rectangular head with serrated or jagged faces. Used for roughing concrete.

Bushed Nipple. A pipe threaded at both ends to connect two pipes of different dimensions.

Bushing. A removable cylindrical lining for an opening used to limit the size of an opening, resist abrasion, or serve as a guide. An electrically insulating lining for a hole to protect a through conductor.

Busway. A rigid assembly consisting of one or more busbars.

Butt Hinge. A hinge consisting of two plates with a removable connecting pin.

Butt Joint. A plain square joint between two members.

Butt, Pile. The large end of a pile. The small end of the pile is called the tip. The pile butt is the end of the pile which the pile driver impacts or "hits" and after completion, the part which the structure is connected to.

Butt Weld. Weld in a joint between two members lying approximately in the same plane.

Butterfly Valve. A valve constructed with a disc that rotates 90 degrees within the valve body.

Buttering. The spreading of a bond coat (followed by a mortar coat, a thin-setting bed mortar, or an organic adhesive) to the backs of ceramic tiles just before they are placed.

Butyl Caulk. Caulking that is made from various synthetic rubbers derived from butanes.

Butyl Membrane. Pliable thin sheets or layers made from synthetic rubber.

BX Cable. Type of indoor wiring consisting of two or more insulated wires protected by a wound, galvanized steel strip cover. The metal winding forms a flexible tube.

C

C-Clamp. A clamp in the shape of a "C" with jaw capacities ranging from 1 to 8 inches used for the securing of wood or metal pieces in a fixed position and for temporary assemblies.

CO₂ Extinguisher. A wheeled or hand-held portable apparatus to put out small fires by ejecting carbon dioxide.

Cabinet. Case, box, or piece of furniture with sets of drawers or shelves, with doors, primarily used for storage.

Cabinet, Base. The floor-mounted cabinets in a room that a counter, sink or appliance is installed upon.

Cable. A thin, flexible line which carries only tensile forces. A bundle of two or more electrical conductors.

Cable Bus. An assembly of insulated cables.

Cable, BX. A type of indoor wiring consisting of two or more insulated wires protected by a wound, galvanized steel strip cover. The metal winding forms a flexible tube. It offers protection similar to rigid conduit.

Cable, Coaxial. A cable to transmit telephone, television and computer signals. A cable that has a tube of conducting material surrounding a central conductor.

Cable, Communication. A cable to transmit telephone, television and computer signals that are transmitted through cable lines.

Cable Tray. Open track for support of insulated cables.

Cage Ladder. A vertical device with rungs for climbing that has, for safety, a surrounding structure to prevent the climber from falling off.

Caisson. A cylindrical, sitecast concrete foundation that penetrates through unsatisfactory soil to rest upon an underlying stratum of rock or satisfactory soil. A type of drilled or augured piling.

Camber. A deflection that is intentionally built into a structural element or form (usually a beam) to improve appearance or to nullify and offset the deflection of the beam under the effects of loads, shrinkage and creep. A radius built into a given member to compensate for deflection. A light initial curvature in a beam or slab.

Cant Strip. A strip of material, usually treated wood or fiber, with a sloping face used to ease the transition from a horizontal to a vertical surface at the edge of a flat roof. This prevents the roofing material from abruptly stopping at the parapet wall and also helps prevent leakage at that juncture.

Cantilever. A structural shape (beam, truss, or slab) that extends beyond its last point of support.

Cap. A trim tile with a convex radius on one edge. Used for finishing the top of a wainscot or for turning an outside corner, a bullnose.

Cap, Pile. A structural member usually fastened to, and placed on the top of a slender timber, concrete or structural element. Used to transmit loads into the pile or group of piles and to connect them.

Capacitor. A device which introduces capacitance into an electric circuit.

Carbon Steel. Low carbon or mild steel.

Carborundum. Very hard material used for various abrasive devices.

Carpentry, Rough. The preliminary framing, boxing and sheeting of a wood frame building.

Carpet Tile. Carpet that comes in sheets or small squares and is installed by the use of adhesives.

Carrier Channel. The main supporting metal members used in the construction of suspended ceilings.

Casement. A window in which the sash opens with hinges and pivots on an axis along the vertical line of the frame.

Casework. Assembled cabinetry or millwork.

Casing. The wood finish pieces surrounding the frame of a window or door, or the finished lumber around a post or beam. A cylindrical steel tube used to line a drilled or driven hole such as a well or caisson.

Cast Iron. Iron with a high carbon content, which cannot, because of the percentage of carbon, be classified as steel.

Cast Iron Pipe. Pipe that is manufactured with an alloy of iron, carbon and silicon that is cast in a mold and is more easily fusible as less corrosive than steel.

Cast-In-Place. Mortar or concrete which is deposited in the place where it is required to harden as part of the structure, as opposed to precast concrete.

Cast-In-Place Concrete. Concrete that is poured in its intended location at a site.

Casting, Solid. Forming castings by introducing a body slip into a porous mold which usually consists of two major sections, one section forming the contour of the inside of the ware and allowing a solid cast to form between the two mold faces.

Catch Basin. Formed pan, usually constructed of masonry, which collects run-off and debris. The basin usually includes a drain connected to plumbing or stormwater system.

Caulking. Waterproof material used to stop up and fill seams to make watertight.

Caveat Emptor. "Let the buyer beware."

Cavity Wall. A masonry wall that includes a continuous airspace between its outermost wythe and the remainder of the wall.

CDX. A grading system mark for plywood which means: grade C and D, exterior glue.

Cedar Shingle. A thin piece of cedar wood with one end thicker than the other for laying in overlapping rows as a covering for a roof or the sides of a building or structure.

Cedar Siding. Boards milled from cedar wood, used for the finish covering on the exterior walls of a building or structure. Used for its resistance to moisture and aging.

Ceiling and Wall. A type of building framing construction.

Ceiling Diffuser. A mechanical device through which warm or cold air is blown into an enclosure. Its purpose is to distribute conditioned air.

Ceiling Framing. Wood or metal pieces which form the rough framing of ceilings.

Ceiling Joist. The horizontal members in a building or structure to which the ceiling material (plaster, drywall etc.) is fastened.

Ceiling Lath. Sheets of expanded metal, gypsum or in older structures, wood slats (1-1/4 inches wide, 3/8 inches thick), which are attached to a ceiling to provide a surface for plastering.

Ceiling Plenum. Space below the flooring and above the suspended ceiling that accommodates the mechanical and electrical equipment and that is used as part of the air distribution system.

Cell. A void space having a gross cross-sectional area greater than 1-1/2 square inches.

Cement. Usually refers to portland cement which when mixed with sand, gravel, and water forms concrete. Generally, cement is an adhesive; specifically, it is that type of adhesive which sets by virtue of a chemical reaction.

Cement Grout. A cementitious mixture of portland cement, sand or other ingredients and water which produces a uniform paste used to fill joints and cavities between masonry units.

Cement, Mortar. A mixture of cement, lime, sand, or other aggregates, and water, used for plastering over masonry or to lay block, brick or tile.

Cement, Plaster. Plaster used on exterior surfaces or in damp areas. Plaster having Portland Cement as a binder.

Cement, Portland. A hydraulic cement produced by pulverizing clinker consisting essentially of hydraulic calcium silicates, and usually containing one or more of the forms of calcium sulfate as an interground addition. The most common type of cement used in construction.

Cement, Masonry. A hydraulic cement for use in mortars for masonry construction, containing one or more of the following materials: portland cement, portland blast-furnace, slag cement, portland-pozzolan cement, natural cement, slag cement or hydraulic lime; and in addition usually containing one or more materials such as hydrated lime, limestone, chalk, calcereous shell, talc, slag, or clay, as prepared for this purpose.

Cementitious. Having cementing properties; usually used with reference to inorganic substances, such as portland cement and lime.

Cementitious Topping. A compound that is capable of setting like concrete and is applied on a concrete base to form a floor surface.

Centering. Temporary formwork for an arch, dome, vault, or other overhead surface.

Centigrade. A scale of temperature which features 0 degrees and 100 degrees as the freezing and boiling points of water respectively. To convert centigrade to Fahrenheit multiply by 1.8 and add 32, e.g. (100 degrees x 1.8)+32 = 212 degrees F.

Central System. A system of conditioning air supplied to various areas or space, serviced by the same source of heat or cooling. All equipment in central systems, except air-cool condenser, evaporative condenser, and cooling towers, are indoors.

Centrifugal Pump. A pump which draws water into the center of a high speed impeller and forces the fluid outward with velocity and pressure.

Centroid. The center of the mass of an object.

Ceramic Tile. A ceramic surfacing unit, usually relatively thin in relation to facial area, made from clay or a mixture of clay; and other ceramic material, called the body of the tile, having either a "glazed" or "unglazed" face and fired above red heat in the course of manufacture to a temperature sufficiently high to produce specific physical properties and characteristics.

Certificate for Payment. A written document forwarded to the general contractor by the architect, engineer, or owner approving payment for work completed.

Certificate of Insurance. A certificate provided by the general contractor verifying that he has obtained the required insurance for the project.

Certificate of Substantial Completion. A written document forwarded to the general contractor by the architect, engineer, or owner indicating that the project is substantially complete. This document initiates the time period for the final payment to the contractor.

CFM. Cubic feet per minute.

Chain Trencher. A self-propelled machine with blades attached to a continuous chain, used to excavate trenches.

Chair. A device used to support reinforcing bars.

Chair, Reinforcing. Metal supports made of fabricated wire, made to hold reinforcing steel in place until concrete is poured.

Chalk Line. Usually cotton cord coated with colored chalk. The cord is snapped to mark straight lines for layout purposes.

Chamfer. A beveled surface cut on the edge of a piece of wood. A strip of wood or other substance cut on an angle and placed in concrete formwork to create a beveled finished surface on the corners of the final concrete shape (beam, column, etc.).

Change Order. An order to change the work to be performed under a construction contract, usually given by an owner to a prime contractor or a by prime contractor to a subcontractor. A revision in the contract documents after the execution of the owner-contractor contract.

Channel. A U-shaped steel or aluminum section shaped like a rectangular box with one side removed.

Channel Furring. A formed sheet metal furring strip.

Chattel. Moveable personal property.

Checking. Short shallow cracks on the surface. When checking refers to wood it is due to shrinkage.

Cherry Veneer. A thin layer of cherry wood used as a finished surface material.

Chestnut Veneer. A thin layer of chestnut wood used as a finished surface material.

Chicken Wire. Thin, woven wire mounted on an exterior wall as a base for stucco plaster to cling to, often backed by paper.

Chilled Water System. A cooling system in which the entire refrigeration cycle occurs within a single piece of equipment. Water is used to bring the heat from the space to the evaporator section of the chiller, and water is also used to carry the heat from the condenser to the outside.

Chimney. A vertical, noncombustible, structure with one or more flues to carry smoke and other gases of combustion into the atmosphere.

Chimney Brick. Brick, chosen for the specific use of the construction of chimneys because of its ability to withstand high temperatures without cracking.

Chimney Flue. A channel or shaft in a chimney for conveying smoke and exhaust gases.

Chipping Hammer. The chipping hammer is a lightweight hammer that comes in a variety of sizes. The head and back can be capped with tungsten carbide for durability. It is used by the mason to chip excess material from the backs and edges of block, brick, stone, or tile.

Chock. Heavy timber or wooden block, fitted under tires or wheels to prevent movement.

Chord. One of the main members of a truss braced by web members of the truss. Perimeter member of a building or structure which resists lateral forces.

Chrome Plated Fitting. A fitting that is plated with an alloy of chromium.

Circuit Breaker. An overcurrent protection device.

Cladding Panel. A panel applied to a structure to provide durability, weathering, corrosion and impact resistance.

Clapboards. A special form of outside covering of a house; siding. Usually made of wood, clapboards are tapered with the thicker side of the board exposed and the thinner side lapped under the next course of boards.

Clay. A natural mineral aggregate, consisting essentially of hydrous aluminum silicates; it is plastic when sufficiently wetted, rigid when dried, and vitrified when fired to a sufficiently high temperature.

Clay Brick. A type of brick manufactured from fine-grained materials mainly from hydrated silicates of aluminum. It is soft and cohesive when moist, but becomes hard when baked or fired.

Clay Pipe. Pipe used for drainage systems and sanitary sewers made of earthenware and glazed to eliminate porosity.

Clay Tile. Earthenware tile that is fired and is used on roofs. Known as quarry tile when used for flooring.

Cleaning Masonry. The final removal of excess grout, excess concrete, etc. from an exterior masonry structure.

Cleanout. An opening to the bottom of a space of sufficient size and spacing to allow the removal of debris. In plumbing, a fitting in a pipeline which can be easily accessed to remove foreign objects or provide an opening to insert cleaning type devices.

Clear Dimension, Clear Opening. The dimension between opposing inside faces or walls of an opening or a room.

Cleats. A small block of material (usually wood) which is fastened to a secure surface and used for attachment or as a toe hold or stopping device for another supporting member.

Cleft, Natural. A natural V-shaped channel, space, opening or fissure in a material.

Closer. The last masonry unit laid in a course; a partial masonry unit used at the corner of a course to adjust the joint spacing. A hydraulic device used to close doors.

Closet Flange. The fitting attached to a subfloor onto which the toilet bowl is attached.

Cloud. A defect in the title to real estate or property. When a property has a cloud on it, it is difficult to sell or complete escrow.

CMU. Concrete Masonry Unit (concrete block).

Coated Roof. A roof, usually flat, that has an asphaltic material applied to it to seal against the elements.

Cofferdam. A watertight enclosure from which water is pumped to expose the bottom of a body of water and permit construction.

Cohesive Soil. A soil, such as clay, the particles of which will adhere to one another by means of cohesive and adhesive forces.

Cold Formed Steel. Process of shaping steel without using heat.

Cold Joint. Any point in a tile installation when tile and setting bed have terminated and the surface has lost its plasticity before work is continued. In road construction, a paving joint in which one strip of asphalt is installed at a different time from the other and bonding is not enhanced. The joint formed between hardened concrete and freshly poured concrete.

Collar. A compression ring around a small circular opening.

Collateral. Property in the possession of a creditor to guarantee payment by a debtor.

Column. A structural member used primarily to support axial compression loads and with a height of at least three times its least lateral dimension. An upright structural member acting primarily in compression. A square, rectangular, or cylindrical support for roofs, ceilings, and so forth, composed of base, shaft, and capital.

Column Capital. The uppermost member of a column crowning the shaft and taking the weight of the beam or girder.

Column Footing. Commonly known as individual footing, generally square or rectangular in shape.

Combination Frame. A combination of the principal features of the full and balloon frames.

Combined Footing. A concrete footing which supports two or more columns.

Commissioning. Start-up of a building that includes testing and adjusting HVAC, electrical, plumbing, and other systems to assure proper functioning and adherence to design criteria. Commissioning also includes the instruction of building representatives in the use of the building systems.

Common Bond. Brickwork laid with each five courses of alternating stretchers followed by one course of headers.

Common Enemy Doctrine. The legal doctrine that flood waters are a common enemy and that property owners may fight to protect their property regardless of the damage to neighboring property.

Compact Borrow. Fill acquired from excavation that has been compacted.

Compacted Concrete. Freshly pored concrete that has been packed tighter by vibration, tamping or a combination of both to remove voids.

Compaction. The process whereby the volume of freshly placed material is reduced or flattened by vibration or tamping, or some combination of these.

Comparative Negligence. The legal doctrine that wrongdoers should pay damages proportional to their fault.

Compensatory Damages. An amount calculated to compensate a party for economic loss caused by the wrongful act of another.

Composite Beam. A beam that is composed of two different materials; for example, a wood and a steel beam, or a steel beam and concrete slab, in which the two act as one.

Compression. Force which tends to crush adjacent particles of a material together and cause overall shortening in the direction of its action. Stress which tends to shorten a member.

Compression Fitting. Bends, couplings, crosses, elbows, tees, unions, etc., which use a force when connecting that pushes together and squeezes a metal or rubber gasket.

Compression Strength. The ability of a structural material to withstand compression forces. The measured maximum resistance of a concrete or mortar specimen to axial loading; expressed as force per unit cross-sectional area; or the specified resistance used in design calculations, in the U.S. customary units of measure expressed in pounds per square inch (psi).

Compressor. A machine that compresses gases or air.

Concave Joint. A mortar joint tooled into a curved, indented profile.

Concealed Grid. A suspended ceiling framework that is completely hidden by the tiles or panels it supports.

Concealed Z Bar. A hidden z-shaped bar that is used as a wall tie.

Concentrated Load. A load which acts at one point or small area of a structure or member.

Concrete. A composite material which consists essentially of a binding medium within which is embedded particles or fragments of aggregate; in portland cement concrete, the binder is a mixture of portland cement and water. A mixture of cement, water, fine aggregate (sand), and coarse aggregate (gravel). An artificial building material made by mixing cement and sand with gravel, broken stone, or other aggregate, and sufficient water to cause the cement to set and bind the entire mass.

Concrete Accessory. An implement or device used in the formwork, pouring, spreading, finishing, etc. of concrete surfaces.

Concrete Admixture. A substance added to concrete to aid in imparting color, control workability, help in waterproofing, control setting and to entrain air.

Concrete Block. A hollow concrete masonry unit.

Concrete Cutting. A large saw with a specific blade used to score or cut concrete.

Concrete Dowel. A pin of reinforcing embedded in concrete to strengthen two pieces where they join or to create a place where other pieces can be fastened to it.

Concrete, Field. Concrete delivered or mixed, placed and cured on the job site.

Concrete Finish. The act or process of the final compaction and finishing operations of curing concrete.

Concrete, Green. Concrete which has set but which has not appreciably cured or hardened.

Concrete Masonry Unit (CMU). A block of hardened concrete, with or without hollow cores, designed to be laid in the same manner as a brick; a concrete block.

Concrete Pipe. Pipe manufactured from concrete. The manufacturing is done in a plant under controlled conditions. Concrete pipe is usually used for drainage but may be used for sanitary sewers also.

Concrete Placement. The placing and finishing of concrete during a continuous operation. Also known as pouring.

Concrete Plank. A solid or hollow-core, flat-beam used for floor or roof decking. Usually precast and prestressed.

Concrete Precast. Cast and cured concrete manufactured in a plant under controlled conditions. Examples are; precast concrete slabs, precast reinforced lintels, beams, columns, piles, parts of walls and floors, etc.

Concrete, Prestressed. Concrete in which internal stresses of such magnitude and distribution are introduced so that the tensile stresses resulting from the structure's loads are counteracted to a desired degree. The stresses are usually developed by inserting tendons (cables) through preformed tubes in the concrete member, at which time the tendon in stressed (tightened) and then grouted in place.

Concrete Pump. An apparatus which forces concrete to the placing position through a pipeline or hose.

Concrete Reinforcement. Steel rods that are embedded in wet concrete to give additional strength.

Concrete Testing. Testing to determine the plasticity or strength of concrete.

Concrete Topping. A rich mixture of fine aggregate concrete used to top concrete floor surfaces for durability, safety and appearance.

Concurrent. The point at which the line of action of several forces meet.

Condenser. A heat exchanger in a refrigeration cycle used to discharge heat to the outside. Three commonly used types of condensers can be classified as water cooled, air cooled, and evaporative. Condenser water is normally circulated through a cooling tower through which heat is distributed to the atmosphere. A heat exchanger in a refrigeration cycle used to discharge heat to the outside.

Conductive Floor. Flooring material specifically designed to prevent electrostatic buildup and sparking.

Conductor. A substance or body capable of transmitting electricity, heat or sound.

Conductors. Pipes for conducting water from a roof to the ground or to a receptacle or drain; downspout. Any electrical wire used to convey electricity.

Conduit. A protective sleeve or pipe commonly used for individual electrical conductors.

Connection. The union, or joint, of two or more distinct elements. In a structure, the connection itself often becomes an entity. Thus, the actions of the parts on each other may be visualized in terms of their action on the connection.

Connector, Compression. A connecting device which when attaching uses a force that pushes together and squeezes.

Connector Set Screw. A screw on a connector fitting that when tightened connects two components together.

Consideration. A benefit (or money) coming from a promisee to a promisor in exchange for the promisor's agreement to perform an act.

Construction Change Directive. A document that directs a change in the work of the project, prepared by the architect and signed by the owner and architect. This change can adjust the contract sum and/or contract time. This document may be used in the absence of total agreement on the terms of a change order.

Continuous Beam. A beam that is supported by more than two supports.

Continuous Footing. A concrete footing that supports a wall or two or more columns. The footing can vary in width and depth. Sometimes called a strip footing.

Continuous High Chair. A rigid wire device used to hold steel reinforcements off the bottom of the slab. As opposed to a single chair, the continuous chair is manufactured in strips.

Contraction Joint. Formed, sawed, or tooled groove in a structure to create a weakened plane and regulate the location or cracking resulting from the dimensional change of different parts of the structure.

Control Joint. An intentional linear discontinuity in a structure or component, designed to form a plane of weakness where cracking or movement can occur in response to various forces so as to minimize or eliminate cracking elsewhere in the structure.

Controls. A mechanism used to regulate or guide the operation of a machine, apparatus, or HVAC system.

Convector. A heat exchange device that uses the heat in steam, hot water, or an electric resistance element to warm the air in a room; often called, inaccurately, a radiator.

Conveying System. A device used to move material from one place to another. A broad term which includes elevators, escalators, moving walks, dumbwaiters, etc..

Coping. The material or units used to form a cap or finish on top of a wall, pier, pilaster, or chimney. A protective cap at the top of a masonry wall.

Copper Braid. Three or more strands of copper intertwined to form a regular diagonal pattern down its length.

Copper Cable. Copper wires conducting power.

Copper Conductor. A copper wire or cable through which power runs from a source to a destination.

Copper Flashing. Sheets of copper that cover open joints on exterior construction to prevent water leakage.

Copper Piping. Pipe and tubing manufactured of copper, classified as Type K, L or M. Type K being the thickest walled, Type M, the thinnest walled.

Corbel. A projection from the face of a beam, girder, column, or wall used as a beam seat or a decoration. A spanning device in which masonry units in successive courses are cantilevered slightly over one another; a projecting bracket of masonry or concrete. A masonry unit such as brick or stone which projects beyond the unit below.

Core Drilling. The process of drilling which extracts a cylindrical sample of concrete, rock or soil. Sometimes used to install pipe or conduit in or through an existing concrete or masonry wall.

Coring Concrete. To drill concrete for samples or to create a void in concrete masonry for conduits or pipe.

Corner Bead. A metal or plastic strip used to form a neat, durable edge at an outside corner of two walls of plaster or gypsum board.

Corner Cabinet. A cabinet wall unit that extends down two walls from the inside corner point.

Corner Guard. Type of molding that is mounted on outside corners in a room or space for finishing and for the protection of the corner from damage.

Cornice. The exterior detail at the meeting of a wall and a roof overhang; a decorative molding at the intersection of a wall and a ceiling. The molded projection which finishes the top of the wall of a building.

Corrosion. The eating and wearing away by chemical action (pitting, rusting).

Cost Estimate. Determined by one of the following methods: a. Area and volume method: The estimate is based on a cost per square foot of the building and/or a cost per cubic foot of the building. b. Unit use: The estimate is based on the cost of one unit multiplied by the number of units in the project. For example, in a hospital, the cost of one patient unit multiplied by the number of patient units in the project. c. In-place unit: The estimate is based on the cost in-place of a unit, such as doors, cubic yards of concrete etc..

Counterflashing. An inverted L-shaped metal strip built into a wall to overlap flashing and make a roof or wall watertight. A flashing turned down from above to overlap another flashing turned up from below so as to shed water.

Countersunk Plug. A wooden peg used to fill a drilled hole in a wooden surface.

Couple. Where a pair for forces of equal magnitude acting in parallel but opposite directions are capable of causing rotation.

Course. A horizontal layer of masonry units, one unit high; a horizontal line of shingles.

Covenant. A promise.

Coursed. Laid in courses with straight bed joints.

Cove. A trim tile unit having one edge with a concave radius. A cove is used to form a junction between the bottom wall course and the floor or to form an inside corner.

Cove Base. A flexible strip of plastic or synthetic rubber used to finish the junction between floor and wall.

Cove Molding. Molding that is concave-shaped. Used to cover interior angles, such as that between the ceiling and a wall.

Cover. In reinforced concrete; the depth or thickness of concrete between the outer-most reinforcing bars and the surface.

Cover, Manhole. A heavy, usually round, steel or iron cover used to gain access to underground work through a manhole.

Coverage. A measure of the amount of material required to cover a given surface.

Crane. A machine for raising, shifting and lowering heavy weights by means of a projecting swinging arm or with the hoisting apparatus supported on an overhead track.

Crane, Traveling. A crane which is connected to tires or tracks and can be readily moved.

Crawl Space Insulation. Insulation that is applied between the first floor joists just above the ground.

Creosote. A brownish oily liquid obtained by distillation of coal tar, used as a wood preservative.

Crew Trailer. A trailer provided on a job site for use by employees.

Cripple Stud. A wood wall framing member that is shorter than full-length studs because it is interrupted by a header or sill. A spacer between horizontal wood members.

Cross Brace. Bracing with two intersecting diagonals. Slender diagonal member within framed wall or partition, to support wall or partition and to withstand structural loads imposed by wind and suction loads, building loads, movement and deflection of structure.

Crown Molding. A decorative type of molding used to make ceiling to wall transitions.

Crusher Run. Gravel, rock, boulders, or blasted rock that has been reduced in size by a machine, but has not been sorted for size.

Culvert. A drain pipe or masonry crossing made of concrete, galvanized corrugated metal, aluminum or steel constructed under a road or embankment to provide a waterway.

Curb Inlet. The opening in a curb through which water flows and drains.

Curb Roof. A roof with a slope that is divided into two pitches of each side. Also known as a gambrel roof or mansard roof.

Curing. Maintenance of humidity and temperature of freshly placed concrete during some definite period following placing, casting, or finishing to assure satisfactory hydration of the cementitious materials and proper hardening of the concrete. The hardening of concrete or plaster.

Curling. The distortion of an originally essentially linear or planar member into a curved shape such as the warping of a slab due to creep or to differences in temperature or moisture content in the zones adjacent to its opposite faces.

Curtain Wall. An exterior building wall that is supported, entirely by the frame of the building, rather than being self-supporting or loadbearing.

Curved Wall Form. A curved shaped support for the placing, pouring and curing of concrete in a curved wall system.

Cushion-Edged Tile. Tile on which the facial edges have a distinct curvature that results in a slightly recessed joint.

Cut and Fill. Excavated material removed from one location and used as fill material in another location.

Cylinder Test. A test to determine the compressive strength of concrete.

D

D Load. A constant load that in structures is due to the mass of the members, the supported structure, and permanent attachments or accessories.

Damper. A flap to control or obstruct the flow of air or other gasses; specifically, a metal control flap in the throat of a fireplace, or in an air duct. Controls that vary airflow through an air outlet, inlet, or duct. A damper position may be immovable, manually adjustable, or part of an automated control system.

Dampproofing. The treatment of concrete or mortar to help prevent the passage or absorption of water. Applied with a suitable coating to exposed surfaces or applied with a suitable admixture of treated cement.

Darby. A hand-manipulated straightedge, usually 3 to 8 ft. (1 to 2.5 m) long, used in the leveling operation of the early stage of concrete placement, preceding supplemental floating and finishing. A stiff straightedge of wood or metal used to level the surface of wet plaster.

dB. Decibel

Dead Load. The load due to the vertical weight of all permanent structural and nonstructural elements of a building, such as walls, floors, and roofs.

Deadman. A large or heavy object buried in the ground as an anchor.

Decibel. The ratio of sound pressure to a base level chosen at the threshold of hearing.

Deck. The form on which concrete for a slab is placed, also the floor or roof slab itself.

Decorative Tile. Tile with a ceramic decoration on the surface.

Deed of Trust. Mortgage.

Deed Restrictions. Restrictive covenants often included in a deed which may limit future use(s) of the land, and other restrictions such as height, size, aesthetics, etc., as long as they are not against the public interest.

Deeds. A general term which refers to all documents conveying property, from one person to another.

Default. A material failure to perform the requirements of a contract.

Deflection. A variation in position or shape of a structure or structural element due to effects of loads or volume change, usually measured as a linear deviation from an established plane rather than an angular variation. Displacement or bending of a structural member due to application of external force.

Deformation. A change in the shape of a structure or structural element caused by a load, or force, acting on the structure.

Deformed Reinforcement. Deformed reinforcing bars, bar and rod mats, and deformed wire.

Degree Day Method. A method of computing fuel requirements for HVAC systems.

Delamination. Separation of one layer from another.

Demolish. To raze or tear down a building or structure.

Deposition. The extra-judicial testimony of a witness given under oath; the written record thereof.

Dewatering. To remove water from the ground or excavations with pumps, wellpoints or drainage systems.

Diagonal. Inclined member of a truss or bracing system used for stiffening and/or wind bracing.

Diagonal Bracing. That form of bracing that diagonally connects joints at different levels.

Diaphragm. A thin, usually rectangular or square element of a structure that is capable of resisting lateral forces in its plane, such as a floor or roof.

Diaphragm Pump. A water pump used to continuously remove water from excavations containing mud and small stones.

Diecast Set Screw. A set screw cast by forcing molten metal into a mold.

Diffuser. A device for distributing air in a forced air heating/cooling system. Often flush-mounted on a ceiling, it has slats to direct the conditioned air evenly into the room or space. Components of the ventilation system that distribute and diffuse air to promote air circulation in the occupied space. A lens in a light fixture to diffuse light.

Dimension Lumber. The most common type of framing lumber. For example, lumber which is 2 inches (nominal) thick and commonly is called "2 by 4", "2 by 6", etc.

Direct Burial Conduit. Electric conduit suitable for burial in exterior applications with an outer surface that resists moisture, fungus, and corrosion.

Direct Current (DC). An electrical current that has a constant flow rate induced by a constant voltage.

Direct Examination. Questions directed to a friendly witness.

Direct Expansion (DX). Interior air cooled by directly passing over an evaporator in which refrigerant is expanding from a fixed reference point.

Discovery. A process in which parties to a lawsuit are made to divulge information.

Disinfect Pipe. The act or process of transmitting an antimicrobial agent through a piping system.

Dismissal. The act of dropping a lawsuit.

Displacement. Movement away from a fixed reference point.

Distribution. The movement of fresh concrete to its placement point. The circulation of collected heat to living areas from collectors or storage. The movement of any fluid or current to its end point.

Ditch. A long narrow excavation dug in the earth for foundation walls, sewer systems, irrigation systems, electrical cable lines, utilities, etc.

Dome. An arch rotated about a vertical axis passing through its crown, or highest point.

Domestic Marble. Marble which comes from the country or area where the structure in which it used is built.

Domestic Well. A water well dug for household use.

Door Bumper. Rubber tip devices mounted on walls or baseboards that prevent door knobs from marring walls.

Door Closer. A mechanical device attached to a door to make it close automatically.

Door Coordinator. A device installed on a pair of doors with an overlapping astragal, that permits the proper closing sequence. The door without the overlapping astragal would be closed first.

Door, Flush. Doors that have two flat surfaces. They can come in 3 basic types - solid (wood or metal), solid-core, or hollow-core.

Door, Folding. Two or more hinged door leaves that fold upon each other when opened.

Door Frame. The area which completely surrounds a door, made of wood or metal, where the hinges are attached.

Door Guard. Guard fabricated of steel components to provide protection over interior face of door or prevent damage to glass and/or to prevent intrusion.

Door, Roll-up. A door which raises and rolls into a coiled configuration and lowers on tracks on either side.

Door Trim. Wood or metal finishing work, often ornamental, used for covering the joints between door jambs and plaster walls. The locks, knobs and hinges on a door.

Dope Coat. Neat cement applied to the setting bed.

Dormitory Wardrobe. A closet in student housing where clothes are kept. Commonly a free-standing cabinet made to match the cabinets of the dormitory room.

Double Bullnose. A type of trim with the same convex radius on two opposite sides.

Double Glazing. Two parallel sheets of glass with an airspace between.

Double Headed Nail. A nail with two closely-spaced heads to permit easy removal; widely used in concrete formwork as a temporary fastener.

Double Hung Window. The most common style of operable window. It has two sashes that slide vertically in parallel tracks. A window with two overlapping sashes that slide vertically in tracks.

Double Walled Tank. A tank constructed with two walls for leak protection and to increase strength and stability.

Douglas Fir. A species of framing lumber. A tree which supplies framing lumber because of its straight growth.

Dovetail Anchor Slot. A matching interlocking strip or slot which is used with a dovetail fastener.

Dowel. A steel bar, which extends into two adjoining portions of a concrete construction, as at a joint in a pavement slab, so as to connect the portions and transfer shear loads.

Downspout. The vertical part of a drainage system from a roof which forms a channel or pipe to remove water from the gutters.

Dozer. A term used in the trade for a bulldozer.

Dragline. A bucket attachment for a crane commonly used in a marsh or marine area, that digs soft materials that must be excavated at some distance from the crane, and draws the bucket towards itself using a cable.

Drain. A trench, ditch, or pipe designed to carry away waste water.

Drain Field. A system of trenches filled with sand, gravel or crushed stone and a series of pipes to distribute septic tank effluent into the surrounding soil.

Drawings. Plans or blueprints.

Drawings, Shop. Drawings of specific items of the project provided by subcontractors or fabricators.

Drift. Lateral deflection of a building caused by wind or earthquake loads. The horizontal displacement or movement of structure when subjected to lateral forces.

Drift Pin. A tapered steel shaft used to align bolt holes in steel connections during erection.

Drilled Well. The act or process of using a rotary drill to dig for water.

Drip Edge. A discontinuity or strip installed at roof eaves or over a window or wall component to force adhering drops of water to fall free of the face of the building rather than move farther toward the interior. The projection of a window header or water table to allow the water to drip clear of the side of the building below it.

Drop Panel. A thickening of a two-way concrete structure at the head of a column.

Drum Trap. A plumbing trap shaped like a cylinder.

Dry Air. Air that contains no water vapor.

Dry Pack. Concrete or mortar mixtures deposited and consolidated by dry packing. A low-slump grout tamped into the space in a connection between pre-cast concrete members, or between a steel column and the column base-plate.

Dry Stone Wall. A wall of stone that has been constructed without the use of mortar or concrete in its joints.

Dry Well. A well, usually filled with gravel, to receive rain run-off water, in an area where the ground is not very permeable, for example, in areas of clay soil.

Drywall. An interior facing panel consisting of a gypsum core sandwiched between paper faces, also called gypsum board or plasterboard. Different types are available for standard, fire-resistant, water-resistant, and other applications.

Duct, Bus. A prefabricated unit containing one or more electric conductors, often a metal bar, that serves as a common connection for two or more circuits.

Duct Heater. A heating element in a duct of an air-handling system.

Ductile. The ability of a material to deform under tensile load.

Ductile Iron Pipe. Iron pipe that is manufactured so as to render it able to be flexed. Ductile iron pipe has the non-corrosive qualities of cast iron but is not brittle and has the handling characteristics of steel.

Ductwork, Metal. The rigid sheet material out of which the ducts of an HVAC system are manufactured, commonly, galvanized sheet steel, aluminum, or stainless steel.

Due Process. Notice, and an opportunity to be heard.

Dummy Joint. A joint placed strictly for design, in sidewalks and patios.

Dump Truck. A truck used for transporting and dumping loose materials.

Dumpster. A large, heavy metal container used to hold and haul rubbish.

Duplex Box. An electrical box for wiring switches or duplex receptacles.

Dust Cap. A protective cap placed over a device to protect it from the intrusion of foreign materials.

Dusting. The application of dry portland cement to a wet floor or deck mortar surface. A pure coat is thus formed by suction of the dry cement.

Dutch Door. A door with two separately hinged leaves, one above the other, enabling one to be open while the other stays shut.

DWV Fitting. A fitting that is either soldered or compressed on lengths of pipe to connect them in a drain waste vent plumbing system.

Dynamite. A type of explosive; a slang term used by tilesetters when referring to a mortar accelerator.

E

E. Represents the modulus of elasticity.

Earthquake. A term used to describe movement in the earth's crust that produces horizontal and vertical movement on the ground surface.

Earthwork. An embankment or other construction made of earth. Any work involving movement or use of soil and other earthen material.

Easement. A right to utilize real property owned by another. An interest in land owned by another that entitles its holder to a specific limited use.

Eaves. The horizontal edge at the low side of a sloping roof.

Edge, Drip. A discontinuity or strip installed at roof eaves or over a window or wall component to force adhering drops of water to fall free of the face of the building rather than move farther toward the interior. The projection of a window header or water table to allow the water to drip clear of the side of the building below it.

Edge Form. A forming member used to limit the horizontal spread of fresh concrete on flat surfaces.

Effective Depth of Section. The distance from the extreme compression fiber to the centroid of tension compression.

Efflorescence. The residue deposited on the surface of a material by the crystallization of soluble salts. A powdery deposit on the face of a structure of masonry or concrete, caused by the leaching of chemical salts by water migrating from within the structure to the surface.

Elastic. Able to return to its original size and shape after removal of stress.

Elastomeric/Plastomeric Membrane. A rubber-like sheet material used as a roof covering.

Elbow. A pipe fitting which joins two pipes at 90 degree angles.

Electric Boiler. A tank in which water is heated or hot water is stored, controlled by an electric current.

Electric Furnace. An enclosed structure in which heat is produced, controlled by electric current.

Electric Heating. A system that imparts heat or holds something to be heated, controlled by electric current.

Electrical. Relating to, or operated by electric current.

Electrical Fee. The amount of money charged for the inspection or installation for the electrical wiring work in a building or structure.

Elevated Floor. A floor system not supported by a subgrade.

Elevated Slab. A roof slab or floor supported by structural members.

Elevated Slab Formwork. The system of support for a freshly poured or placed concrete elevated slab.

Elevated Stairs. A stair system not supported by the subgrade.

Elevator. A cable or chain hoist conveying system used for raising material or passengers in a cab, cage, or platform.

Embedment Length. The length of embedded reinforcement provided beyond a critical section.

Embezzlement. Theft of property which has become a possession of the thief.

EMT Conduit. "Electrical Metallic Tubing" or "thin wall". For running of electrical wires, that is not threaded, easier to handle than "rigid", and installed more rapidly because of the type of non-threaded fittings used with it.

Enamel Paint. A paint which dries to a hard gloss or semi-gloss smooth finish.

Encased Burial Conduit. Metal or plastic conduit EB for outdoor wiring with type TW wires encased, or type UF (underground feeder) cable.

Encroachment. Personal property of one person intruding upon real estate owned by another.

Encumbrance. A charge against real property.

End Bell. Pipe with a wide opening at one end, commonly cast iron.

End Nail. To drive a nail through one piece of lumber and into the end grain of another.

Engineer. A person who uses the science by which the properties of matter and the sources of energy in nature are made useful to man.

Engineered Fill. Earth compacted into place in such a way that it has predictable physical properties, based on laboratory tests and specified, supervised, installation procedures.

Entraining Agent. A substance added to concrete, mortar or cement that produces air bubbles during mixing, making it easier to work with and increasing its resistance to frost and freezing.

Epoxy Adhesive. A two-part adhesive system employing epoxy resin and epoxy hardener used for bonding of ceramic tile to back-up materials.

Epoxy Concrete. Concrete with added adhesive resin to aid in binding.

Epoxy Resin. An epoxy composition used as a chemical-resistant setting adhesive or chemical-resistant grout.

Equilibrium. A state of rest due to balanced forces. State of being in balance; implies no tendency to change.

Equipment, Architectural. The implements or apparatus installed during the construction or alteration of a structure. Architectural equipment is usually connected to the structure and either plumbed or wired to the building's services.

Equipment Mobilization. The assembly and movement of equipment to a jobsite.

Equipment Pad. A thick slab-type stone or precast concrete block placed under mechanical devices to spread the weight and load of the machinery evenly and to prevent excessive vibration.

Errors and Omissions Insurance. Insurance obtained by the architect as protection against errors and omissions in the preparation of the contract documents.

Escalator. A power-driven set of stairs arranged like an endless belt that ascend or descend continuously.

Escrow. A neutral facilitator who follows the instructions of parties to a transaction.

Estimate. A prediction of the cost of performing work. A value judgement based on experience. An approximation of construction costs.

Estoppel. The doctrine that a person may not contradict his own positive representations.

Etch. The art of producing designs on metal or glass by the use of the corrosive action of an acid. The use of acid to cut lines into metal or remove the surface of concrete.

Evidence. Testimony, documents, and things introduced in a proceeding to support the contentions of the parties.

Ex Parte Proceeding. One party to a dispute appears before a judge when the other is not present.

Excavation. A cavity formed by cutting, digging or scooping.

Exculpatory Clause. A provision in a contract that relieves a party of liability.

Exemplary Damages. Damages awarded to a private litigant and against a defendant to punish the wrongdoing of the defendant.

Exhaust Fan. An electrical powered device to withdraw fumes, dusts, or odors from an enclosure.

Exhaust Hood. Usually a square or rectangular hood housing an exhaust fan to withdraw fumes, dust, or odors from an enclosure.

Expanded Metal Lath. Open mesh cut and drawn from solid sheet of ferrous or non-ferrous metal. Made in various patterns and metal thicknesses with uneven or flattened surface.

Expansion Joint. A joint through tile, mortar, concrete or masonry down to the substrate, intended to allow for gross movement due to thermal stress or material shrinkage.

Exposed Aggregate Finish. A concrete surface in which the coarse aggregate is revealed. A decorative finish for concrete work achieved by removing, generally before the concrete has fully hardened, the outer skin of mortar and exposing the coarse aggregate.

Extra Work. Work performed by a contractor that is not included within the scope of the work defined by the contract documents.

Extruded Tile. A tile or trim unit that is formed when plastic clay mixtures are forced through a pug mill opening (die) of suitable configuration, resulting in a continuous ribbon of formed clay. A wire cutter or similar cutoff device is then used to cut the ribbon into appropriate lengths and widths of tile.

Extrusion. The process of squeezing a material through a shaped orifice to produce a linear element with the desired cross section; an element produced by this process.

F

Fabrication, Metal. The building, construction or manufacture of metal structures or devices.

Facade. An exterior face of a building.

Face Brick. A brick selected on the basis of appearance and durability for use in the exposed surface of a wall.

Factor of Safety. The ratio of ultimate strength to working stress of a material.

Factored Load. The load imposed on a member multiplied by appropriate factors, used to design reinforced concrete members.

Fahrenheit. A scale for registering temperature where freezing is 32 degrees above zero and boiling is 212 degrees.

Fan Coil Unit (FCU). A packaged unit consisting of a heating/cooling coil, fan, and filter, without ductwork, used to serve a space or a group of small spaces.

Fan, Exhaust. An electrical powered device used to withdraw fumes, dusts, or odors from an enclosure.

Fascia. A flat, vertical face member or band at the surface of a building or the edge beam of a bridge, or exposed eaves of a building; often inappropriately called facia. A flat member of a cornice or other finish; generally the board of the cornice to which the gutter is fastened.

Fastener. Generic term for welds, bolts, rivets, screws, and other connecting devices.

Fastener, Insulation. A device used to fasten different types of insulation to floors, walls, roofs, or ceiling surfaces.

Fastener, Pneumatically Driven. Driven pin or threaded stud which is driven into material by use of compressed air.

Fatigue. A structural failure which occurs as the result of a load being applied and removed, or reversed, repeatedly over a long period of time, or a large number of cycles.

Fee, Professional. The amount of money charged by a person or agency, such as designers, architects, carpenters, contractors, etc. for work performed.

Feeder Duct. An enclosure for a group of electrical conductors that runs power from a central, large source to one or more secondary distribution centers.

Fence. A barrier used to prevent escape or intrusion or to mark a boundary, usually made of posts, boards or wire.

Fence Hole. A hole in the ground where the main vertical support is inserted in the construction of a fence or wall.

Fender Pile. Outside row of piles that protects a pier or wharf from damage by ships.

Fertilizing. The act or process of adding a substance (as manure or a chemical mixture) in order to make soil more fertile.

Fiberboard. A prefabricated sheet of compressed wood or plant fibers used for building. A homogeneous panel made from wood or cane fibers.

Fiberglass. The name for products made of or with glass fibers ranging from 5 to 600 hundred-thousandths inch in diameter. Used for making textile fabrics, and for heat or sound insulation.

Fiberglass Reinforced Pipe. A pipe for liquid or gas, fabricated from glass fibers and resins for strength and durability.

Fibrous Concrete. A light concrete made from a fibrous aggregate, like sawdust or asbestos, for increased tensile strength and making it easy to nail.

Fiduciary. A person in a relationship or position of trust. A trusted overseer.

Field Engineer. An engineer who works primarily at the jobsite as opposed to the home office. Commonly represents the owner or agency and often makes small engineering changes at the site to facilitate construction.

Field Tile. An area of tile covering a wall or floor. The field is bordered by tile trim.

Field Welded Truss. A truss fabricated and welded at a jobsite.

Fifty-Fifty. A dry or dampened mixture of one part portland cement and one part extra-fine sand. This mix is used as a filler in the joints of mounted ceramic mosaic tiles to keep them evenly spaced during installation.

Filler, Joint. Compressible material used to fill a joint to prevent the infiltration of debris and to provide support for sealants.

Fillet Weld. A weld at the inside intersection of two metal surfaces that meet at right angles.

Filter. An adjunct to air cleaning device available to serve four purposes: 1. Commercial filter - used to remove visible particles of dust, dirt and debris. 2. Electrostatic filter - to remove microscopic particles such as smoke and haze. 3. Activated charcoal filter - to destroy odors. 4. Ultraviolet lamps or chemicals - to kill bacteria.

Fineness Modulus. A factor obtained by adding the total percentages by weight of an aggregate sample retained on each of a specified series of sieves, and dividing the sum by 100. In the United States the standard sieve sizes are No. 100 (150 μm), No. 30 (600 μm), No. 16 (1.18 mm), No. 8 (2.36 mm) and No. 4 (4.75 mm), and 3/8 in. (9.5 mm), 3/4 in. (19 mm), 1-1/2 in. (38.1 mm), 3 in. (75 mm), and 6 in. (150 mm).

Finger Joint. A glued end connection between two pieces of wood, using an interlocking pattern of deeply cut "fingers". A finger joint creates a large surface for the glue bond, to allow it to develop the full tensile strength of the wood it connects.

Finish. The final primarily visible work on a construction project.

Finish Carpentry. Installation of wood finish and materials to a building, such as doors, moldings, and window trim.

Finish Plaster. The final layer of plaster coating.

Finishing Sheetrock. The taping and sanding of sheetrock seams to make ready for painting or finish.

Fink Truss. A three triangle symmetrical truss, commonly used in supporting large, sloping roofs.

Fire Clay. An earthy or stony mineral aggregate which has as the essential constituent hydrous silicates of aluminum with or without free silica, plastic when sufficiently pulverized and wetted, rigid when subsequently dried, and suitable for use in commercial refractory products.

Fire Damper. A damper that automatically closes a duct or opening upon detection of fire.

Fire Hydrant. A discharge fitting or apparatus with a valve and spout at which water may be drawn from the mains of waterworks. Used primarily in fire-fighting, also used to service water mains and systems.

Fire Rated Brick. Brick that has been tested for fire-resistance and then graded for specific construction uses.

Fire Rated Door. A door which has been given a rating of how long in time it can withstand fire before failure.

Fire Resistant Sheetrock. Sheetrock which has been manufactured with fire-resistant chemicals.

Fire Retardant. Denotes a reduction in flammability and in spread of fire. Applied or pressurized chemical treatment to retard combustion.

Fire Separation Wall. A wall required under the building code to divide two parts of a building as a deterrent to the spread of fire. A fire wall.

Firebrick. A brick made from special clays to withstand very high temperatures, as in a fireplace, furnace, or industrial chimney.

Firecut. A sloping end cut on a wood beam or joist where it enters a masonry wall. The purpose of the firecut is to allow the wood member to rotate out of the wall without prying the wall apart, if the floor or roof structure should burn through in a fire.

Fireproofing. Material applied to an element to insulate it against excessive temperatures in case of fire.

Firestop. A wood baffle used to close an opening between studs or joists in a balloon or platform framing in order to retard the spread of fire through the opening.

Firewall. A wall constructed to prevent or slow down the spread of fire.

Fished Joint. An end butt splice strengthened by pieces nailed on the sides.

Fitting. A device used for connecting.

Fixed End Beam. A beam fixed upon a support which prevents its rotation.

Fixed Window. Glass that is immovably mounted in a wall.

Fixture Carrier. A wall mounted frame for mounting and support of a plumbing fixture.

Fixture, Lighting. An electrical device installed to illuminated an area with light.

Fixture, Plumbing. Plumbing devices such as sinks, water closets, shower/bath units, etc..

Flammable. Capable of being easily ignited.

Flange. A rib or rim on an object for strength, for guiding, or for attachment to another object. A projecting hard ring, ridge, or collar placed on a pipe or shaft to strengthen, prevent sliding, or accommodate attachments.

Flash Cove. A detail in which a sheet of resilient flooring is turned up at the edge and finished against the wall to create an integral baseboard.

Flashing. A thin, continuous sheet of metal, plastic, rubber or waterproof paper used to prevent the passage of water through a joint in a wall, roof or at a chimney. The material used and the process of making watertight the roof intersections and other exposed places on the outside of a structure.

Flat Seam. A sheet metal roofing seam that is formed flat against the surface of the roof.

Flat Slab. A reinforced concrete slab that is designed to span, without any beams or girders, in two directions to supporting columns.

Flat Washer. A washer which goes under a bolt head or a nut to spread the load, prevent loosening, and protect the surface.

Flemish Bond. Brickwork laid with each course consisting of alternating headers and stretchers.

Flexible. Limber, not stiff, the opposite of rigid.

Flexible Conduit. Metallic conduit that is flexible and is put up in coils instead of lengths. May be used in lieu of the rigid type in many dry locations.

Flexible Coupling. A mechanical connection that adapts to misalignment between moving parts.

Flexible Ductwork. Flexible ductwork manufactured in various diameters, made from spiral wire covered in plastic and commonly insulated, for use in the transfer of air in heating, cooling and ventilating systems.

Flexible Wiring. Electrical wiring that permits movement from expansion, contraction, vibration, and/or rotation.

Flitch-Sliced Veneer. A thin sheet of wood cut by passing a block of wood vertically against a long, sharp knife.

Float. A tool or apparatus for smoothing a surface.

Float Trap. A floating device in a plumbing fixture which opens or closes a valve to prevent sewer air gases from escaping back through the fixture.

Floor Box. A flush or raised electrical outlet box providing outlets from conduits in or under a floor.

Floor Tile, Clay. Quarry tile that is fired and used for flooring.

Floor Deck. Sheet steel formed to fluted or ribbed profile to span between supports (usually joists or beams) to support floor system (usually concrete slab) and live loads.

Floor Finish. The stain, paint, wax or polish on a floor.

Floor Joist. A support beam, commonly installed in parallel with other beams to create a structural floor system, after which floor sheathing is fastened.

Floor Masonry. Shaped or molded masonry units such as, stone, ceramic brick, tile or concrete used for finished flooring.

Floor Patch. Material used to repair damage to a floor.

Floor Pedestal. A member, such as a short pier, used as a base for a floor system.

Floor Slab. A reinforced concrete slab on grade or elevated, used as a floor.

Floor, Terrazzo. A high polished floor made from small marble or stone fragments embedded in a matrix.

Floor, Tile. A floor covered by flat units of baked clay, which may be glazed or ornamental.

Flooring. Any type of material used to create a final floor surface.

Flooring, Marble. A floor system using marble as its finish material.

Flue Liner. Heat-resistant firebrick or other fire clay materials that make up the lining of a chimney.

Fluid Applied Roof. A roof coated with an asphalt-based liquid.

Flush. Adjacent surfaces even, or in same plane (with reference to two structural or finish pieces).

Flush Door. A door with a smooth surface and no protrusions.

Fluted Column. Columns with vertical decorative, semi-circular channeled shafts.

Foamglass Insulation. A thermal insulation made by foaming glass with hydrogen sulfide. It is manufactured in the form of block or board and has a low fire hazard rating.

Fog Curing. Storage of concrete in a moist room in which the desired high humidity is achieved by the atomization of fresh water. Application of atomized fresh water to concrete, stucco, mortar, or plaster.

Folded Plate. A roof structure in which strength and stiffness derive from a pleated or folded geometry.

Folding Door. A door with hinged leaves, often provided with a ceiling or floor track.

Folding Stair. A hinged stair, attached to and concealed within the ceiling, which can be raised and lowered.

Food Service Equipment. Appliances, materials, etc., used in the cooking and serving of food in large quantities.

Footing. The widened base of a foundation that spreads a load from the building across a broader area of soil. An enlargement at the lower end of a wall, pier, or column, to distribute the load.

Footing Drain. Drain around the perimeter of a building, at the footings, to drain groundwater or rainwater away from the building.

Force Majeure. An overwhelming, but unanticipated event.

Foreman. An specially trained workman/manager who works with and usually leads a crew or gang.

Form Deck. Thin, corrugated steel decking that serves as permanent formwork for a reinforced concrete deck.

Form Fabric. Welded-wire fabric used to reinforce concrete while it is setting and gaining sufficient strength to be self-supporting.

Form Oil. Oil applied to interior surface of formwork to promote easy release from the concrete when forms are removed.

Form Tie. A steel rod with fasteners on each end, used to hold together the formwork for a concrete wall.

Formica. Brand name and trademark for any of the various laminated plastic products, usually used for surface finish on cabinets or millwork.

Formwork. Temporary structures of wood, steel or plastic that serve to give shape to poured concrete.

Foundation. The portion of a building that has the sole purpose of transmitting structural loads from the building into the earth. That part of a building or wall which supports the superstructure.

Foundation Vent. Opening in foundation wall to provide natural ventilation to foundation crawl spaces.

Frame. A structural system consisting of relatively long, prismatic members fastened together. A rigid frame is one in which the joints can transmit moments as well as forces and which therefore does not require a braced frame for rigidity. The surrounding or enclosing woodwork of windows, doors, etc., the timber skeleton of a building.

Framing. The rough wooden structural skeleton of a building, including interior and exterior walls, floor, roof, and ceilings.

Framing Lumber. Wood members of framing systems which are manufactured by sawing, resawing, passing lengthwise through standard planing machine, crosscutting to length, and matching, but without further manufacturing.

Framing, Timber. Framing by the use of timber for load carrying applications. Timber is larger than dimension lumber (usually over 4 inches thick).

Fraud. A false statement of fact that is designed to deceive.

Friction Connection. Two or more structural steel members clamped together by high-strength bolts with sufficient force that the loads on the members are transmitted between them by friction along their mating surfaces.

Friction Piling. A load-bearing pile that has its supporting capacity from the skin friction between the soil in contact with the pile.

Front End Loader. A tractor or bulldozer with a bucket which operates from the front of the vehicle.

Frost Line. The depth in the earth to which the soil can be expected to freeze during winter.

Full Frame. The old fashioned mortised-and-tenoned frame, in which every joint was mortised and tenoned. Rarely used at the present time.

Furan Mortar. A two-part mortar system of furan resin and furan hardener used for bonding tile to back-up material where chemical resistance of floors is important.

Furnace. A unit which draws in cool air from an occupied space and passes the air through a heating chamber, combustion or electric, and then is returned to the occupied space; a heat system using air as the distribution fluid.

Furnishings. Articles, especially furniture, found in the interior of a structure, generally to increase comfort or utility.

Furring. Wood or metal strips used to build out a surface such as a studded wall. Narrow strips fastened to the walls and ceilings to form a straight surface upon which to lay the lath or other finish.

Furring, Channel. A formed sheet metal furring strip.

Furring, Metal. A length of metal channel attached to a masonry or concrete wall to permit the attachment of finish materials to the wall.

Furring, Wood. Strips of wood applied to a surface to provide fastening base for materials.

Fuse. An overcurrent protection device.

Fused Reducer. A pipe coupling with a larger size at one end that the other and is attached to a length of pipe by welding.

G

Gabion. A metal or wire cage filled with ballast or stone, used in large scale retaining walls.

Gable. The triangular wall beneath the end of a gable roof. The vertical triangular end of a building from the eaves to the apex of the roof.

Gable End Rafter. The last rafter system installed at the gable end of a building.

Gable Louver. A louver system installed at the gable end of a building.

Gage. A tool used by carpenters to strike a line parallel to the edge of a board.

Galvanized. Zinc plated for corrosion protection achieved by hot dipping into molten zinc or by electrolysis.

Galvanized Mesh. Mesh screening that has been galvanized. Commonly used as wire lath, reinforcing, or fencing.

Gambrel. A symmetrical roof with two different pitches or slopes on each side.

Gang Box. Electrical boxes constructed of metal or hard plastic, manufactured with knockout holes to pull wire through to connect outlets, switches and other devices.

Gas Furnace. A heating system that burns gas to produce heat.

Gas Meter. A mechanical device measuring and recording the volume of gas passing a given point.

Gate, Swing. The operable member of a fence system that is hinged for opening and closing.

Gate Valve. A valve utilizing a wedge-shaped gate, which allows fluid flow when the gate is lifted from the seat.

Gauging Plaster. A gypsum plaster formulated for use in combination with finish lime in finish coat plaster.

General Conditions. A written document, supplementing the specifications, which indicates and defines areas of the project relating to other than specific building trades.

Geotextile. Synthetic fabrics used to separate backfill materials for proper drainage. Used in high retaining walls and landscape design.

GFI Breaker. Supplies power, as any breaker does, but also monitors the amount of incoming and outgoing current. Whenever the entering current does not equal the leaving current, indicating current leakage, the GFI instantly opens the circuit. A faster overcurrent protection device than either a fuse or circuit breaker.

Girder. A beam that supports other beams; a very large beam, especially one that is built up from smaller elements. A timber used to support wall beams or joists.

Girt. A beam that supports wall cladding between columns.

Glass. An amorphous inorganic usually transparent or translucent substance consisting of a mixture of silicates or sometimes borates or phosphates formed by fusion of silica or of oxides of boron or phosphorous with a flux and a stabilizer into a mass that cools to a rigid condition without crystallization.

Glass Bead. A narrow strip of plastic, metal, or wood used to hold glass in a sash. Removable trim that holds glass in place.

Glass Block. A hollow masonry unit made of glass.

Glass Pipe. Glass and glass-lined pipe used in process piping and in laboratories.

Glass, Wire. Glass in which a wire mesh was embedded during manufacture.

Glass-Fiber-Reinforced Concrete (GFRC). Concrete with a strengthening admixture of short alkali-resistant glass fiber.

Glazed Block. Concrete blocks with a surface produced by fusing it with a glazing material.

Glazed Brick. Brick or tile with a surface produced by fusing it with a glazing material.

Glazing. The act or process of furnishing or fitting with glass. A transparent or translucent color applied to modify the effect of a painted surface.

Glazing Compound. Any of a number of types of mastic or putty used to bed small lights of glass in a frame.

Globe Valve. A valve with a rounded disc that shuts off the flow when closed, and seats to prevent fluid flow.

Glue Laminated Timber. A timber made up of a large number of small strips of wood glued together.

Glulam. An abbreviation of glue laminated timber.

Grab Bar. Metal or plastic bar attached to a bathroom wall, above a bathtub, near a toilet, or in a shower, to be used as a hand hold.

Gradall. A hydraulic, wheel-mounted backhoe often used with a wide bucket for dressing earth slopes.

Grade. A predetermined degree of slope that a finished floor or ramped surface should have. The horizontal ground level of a building or structure.

Grade (Lumber). The classification of lumber in regard to strength and utility in accordance with the rules of an approved lumber grading agency.

Grade Beam. A reinforced concrete beam that transmits the load from a bearing wall into spaced foundations such as pile caps or caissons.

Grading. The act or process of leveling earth.

Granite Block. A masonry unit consisting of a very hard natural igneous rock used for its firmness and endurance.

Granolithic Topping. A covering layer consisting of an artificial stone of crushed granite and cement.

Grass Cloth. A wall covering manufactured from vegetable fibers, woven and laminated to paper for backing.

Grating. Open grid of metal bars structurally formed.

Grating Frame. Frame, usually metal, to contain floor grating and provide means to anchor to floor construction.

Gravel. Small rock particles resulting from natural disintegration and weathering such as river gravel. Or a loose term for mechanically crushed stone.

Gravel Roof. A roof composed of layers of roofing felt for waterproofing, then sealed with tar or pitch and covered with a layer of gravel to assist in protection from wear and the sun.

Gravel Stop. A metal flange or strip with a vertical lip placed around a built-up roof to prevent loose gravel from falling off the roof.

Gravity Ventilator. A device installed in an opening in a room or building which is activated by air passing through to remove stale air and replace it with fresh air.

Gravity Wall. A retaining wall which depends solely on its weight to resist lateral forces of retained earth.

Grease Trap. A device to trap and retain the grease content of wastewater and sewage.

Green Concrete. Concrete that has not hardened to its design strength but has set initially.

Grille. Component of the ventilation system that promotes air circulation in the occupied space by providing a means to return air. A metal screen or grating that allows for the circulation of air.

Ground. A strip of wood assisting the plasterer in making a straight wall and in giving a place to which the finish trim of the room may be nailed. A strip attached to a wall or ceiling to establish the level to which plaster should be applied.

Ground Fault Circuit Interrupter (GFCI). Supplies power, as any receptacle does, but also monitors the amount of incoming and outgoing current. Whenever the entering current does not equal the leaving current, indicating current leakage, the GFCI instantly opens the circuit. A faster overcurrent protection device than either a fuse or circuit breaker.

Ground Hydrant. A water hydrant for the use in fighting fires, installed in the ground.

Grounding. The act or process of making an electrical connection with the earth. A large conduction body (as the earth) used as a common return for an electric circuit.

Grout. A rich or strong cementitious or chemically setting mix used for filling masonry or tile joints and/or voids. A mixture of portland cement, aggregates, and water, which can be poured or pumped into cavities in concrete or masonry.

Grout Lift. An increment of height that grout is poured.

Grubbing. The act or process of clearing and digging up roots and stumps.

Guarantee. Written or implied assurances for a specific part of the project, or for the project as a total. An undertaking or document stating that a thing will or will not happen.

Guardrail. A safety device used as a barrier to prevent encroachment. In street or highway construction, a barrier to keep vehicles in their lanes. A device for protecting a machine part or the operator of a machine.

Gunite. A term used to describe a concrete material applied by pumping through a hose.

Gusset Plate. A flat steel plate to which the chords of a truss are connected at a joint; a stiffener plate.

Gutter. A channel to collect rainwater and snow melt at the eaves of a roof.

Guy Cable. Cable anchored at one end and supporting or stabilizing an object at other end.

Guy Rod. A metal rod with a cable or rope attached, leading to an object to support and stabilize it.

Gypsum Board. An interior facing panel consisting of a gypsum core sandwiched between paper faces. Also called drywall or plasterboard.

Gypsum Lath. Small sheets of gypsum board manufactured specifically for use as a plaster base.

H

H Beam. A steel beam which in cross section resembles the letter "H". Commonly used in earthwork as a retaining structure or piling.

Half Round Molding. An ornamental strip having one flat side and one rounded side.

Halide Fixture. An electric-discharge light fixture that makes its light from metal vapor such as sodium or mercury.

Halon System. A system using halon gas for the fire protection of water sensitive equipment.

Hand Excavation. The act or process of digging out earth using hand tools.

Handhole. Access hole used for repair and cleaning.

Handicapped Plumbing. Plumbing devices and layouts specifically designed for use by disabled individuals.

Handling, Air. Single or variable speed fans pushing air over hot or cold coils, through dampers and ducts to heat or cool a building or structure.

Handrail. Member which is normally grasped by hand for support.

Handrail, Pipe. Metal pipe member which is normally grasped by hand for support.

Handsplit Shingle. A shingle made by splitting a block of wood, usually cedar or redwood, along its grain and thereby creating a shingle which may be used for roofing or siding. Another term used may be "shake shingles".

Hanger, Joist. A metal stirrup that supports the ends of joists so that they are flush with the girder.

Hanger Rod. A rod for connecting pipe, gutters, or ceiling framework to a support.

Hardboard. A very dense panel product, usually with at least one smooth face, made of highly compressed wood fibers. Flat sheet material of fibers consolidated under heat and pressure in hot press.

Hardware. A general term for metal or plastic fittings used in or on a building and its parts.

Hasp Assembly. A fastener for a door or lid consisting of a hinged metal strap that fits over a loop and is secured by a pin or padlock.

Hatch. A hinged or removable cover in a floor or roof which permits ventilation or the passage of persons or objects.

Hawk. A metal square with a wooden handle at the center, used to temporarily hold mortar or plaster by a plasterer or tilesetter.

Hazardous Waste. A material or substance characterized by a propensity to be unhealthy or dangerous.

Head Joint. The vertical layer of mortar between ends of masonry units.

Head, Pop-Up. A watering device in an irrigation or sprinkling system that pops up when the system is charged with water.

Header. A lintel. A joist that supports other joists. A short joist into which the common joists are framed around or over an opening. A structural support over an opening.

Headroom. The clear space between the floor line and ceiling, as in a stairway.

Headwall. A wall, usually constructed of concrete or masonry, that is placed at the outlet side of a drain or culvert to protect fill from scouring, undermining or to divert flow.

Hearing. A proceeding conducted by a judge or arbitrator who receives evidence about a dispute.

Hearsay. Evidence of an event that the witness did not personally perceive.

Heartwood. The wood cells nearer the center of a tree trunk.

Heat of Hydration. The thermal energy given off by concrete, masonry, or gypsum as it cures.

Heat Pump. A unit that supplies heating or cooling to an interior space by either absorbing heat from or rejecting heat to the outside.

Heat Transfer. Heat flow from an area of high temperature to an area of low temperature ending in an equalization of the two areas.

Heater, Unit. A device for heating a space without the use of ductwork.

Heavy Timber. Construction requiring noncombustible exterior walls with a minimal fire-resistance rating of two-hours, laminated or solid interior members, heavy plank or laminated wood floors and roofs.

Heel, Rafter. The end or foot that rests on the wall plate.

HEPA. High efficiency particulate arrestance (filters).

High Chair Reinforcing. A term used in the trade describing a chair-shaped device used to hold steel reinforcement off of the slab while the concrete is being poured.

High Lift Grouting. A method of construction in which concrete block units may be laid the entire height of the wall before grouting. This method of construction requires special inspection.

Highbay Lighting. A lighting system located high above work or floor level.

Hinge. A joint fixing the relative position of the ends of two or more structural members, but permitting their relative rotation. A piece of door hardware that permits the opening and closing of a door by joining the door to the jamb with a flexible device.

Hip Roof. A roof consisting of four sloping planes that intersect to form a pyramidal or elongated pyramid shape. A roof which slopes up toward the center from all sides, necessitating a hip rafter at each corner.

Hod. A portable trough for carrying mortar, bricks, etc., fixed crosswise on top of a pole and carried on the shoulder.

Hollow Block. Concrete blocks that can be filled with insulation or reinforced.

Hood, Exhaust. Usually square or rectangular hood housing an exhaust fan to withdraw fumes, dusts, or odors from an enclosure.

Hook. A semicircular bend in the end of a reinforcing bar.

Hopped-Up Mud. Mortar mixed with an accelerator.

Hopper Window. A window whose sash pivots on an axis along the sill, and that opens by tilting toward the interior of the building.

Horizontal Damper. A damper that disperses forces in a horizontal plane.

Horizontal Siding. Linear horizontal material, usually overlapping, used as exterior surface or cladding for exterior framed wall to provide protection from exterior elements.

Horizontal Wood Siding. Linear horizontal wood material, usually overlapping, used as exterior surface or cladding for exterior framed wall to provide protection from exterior elements.

Hose Bibb. An outdoor water faucet, usually at sill height, used as a hose connection.

Hot Mud or Hot Stuff. Mortar mixed with an accelerator.

House Trap. Located at the point at which the house drain leaves the building, designed to hold a quantity of water that prevents gasses from the sewer system from entering the building.

Housed Joint. A joint in which a piece is grooved to receive the piece which is to form the other part of the joint.

HP Sodium Lamp. A high-pressure (HP) sodium vapor lamp that produces a wide-spectrum yellow light.

Hub Union. A pipe fitting used to join two pipes without turning either pipe.

Humidification. The addition of moisture to air, thereby increasing the latent heat.

HVAC. Heating, ventilation, and air-conditioning system.

Hydrant. A connection with a valve to a water main, used for the dispensation and delivery of water.

Hydrant, Wall. A connection to a water main cut through and mounted on a wall.

Hydrate. A chemical combination of water with another compound or an element.

Hydrated Lime. Calcium hydroxide, a dry powder obtained by treating quicklime with water.

Hydraulic Elevator. A lifting device powered by pressured fluid.

Hydraulic Excavator. A digging machine powered by pressured fluid.

Hydraulic Lift. A elevating device powered by pressured fluid.

Hydronic Heating System. A system that circulates heated water through convectors to heat a building or space.

Hydrostatic Pressure. Pressure exerted by standing water.

I

I. Represents the moment of inertia of a member.

I Beam. An obsolete term; an American Standard designation for a particular section of hot-rolled steel which in cross section is shaped like a capital "I".

IAQ. Indoor air quality.

Ice Dam. An ice obstruction along the eaves of a roof caused by the refreezing of water emanating from melting snow on the roof surface above.

Illegal. Contrary to or against the law.

IMC (Conduit). Intermediate Metal Conduit.

Immune. Not subject to legal process.

Impact Barrier. A barrier constructed resist dynamic loading on a surface.

Impact Insulation Class (IIC). A single figure rating that provides an estimate of the sound insulating performance of a floor-ceiling assembly.

Impervious. Not letting water or moisture pass through or be absorbed in. The degree of vitrification evidenced visually by complete resistance to dye penetration. The term impervious generally signifies zero absorption, except for floor and wall tile which are considered "impervious" up to 0.5 percent water absorption.

Implied Contract. A contract that is not in words.

Implied Covenant. A promise that is not expressed in a contract, but that is implied from the surrounding circumstances.

Implied Indemnity. An obligation to indemnify that arises not from the words of a contract, but from the circumstances of the parties.

Implied Warranty. A promise, usually related to a the quality or serviceability of goods, that is not in words but is implied from the circumstances of a sale.

Impossibility. A doctrine of contract law that excuses performance that becomes physically impossible.

Incandescent Fixture. An electric light fixture in which the lamp or bulb has a filament that gives off light when heated to incandescence by an electric current.

Incandescent Lamp. A device that produces light through an electrically heated filament.

Indemnity. A promise to hold a person harmless from liability or loss.

Independent Contractor. A person who, in performing services for another, is responsible only for the final result, and is not subject to control as to the methods used to achieve that result.

Indeterminate. Usually short for statically indeterminate

Indicator Compounds. Chemical compounds, such as carbon dioxide, whose presence at certain concentrations may be used to estimate certain building conditions (e.g., airflow).

Indirect Luminaire. A lighting fixture that distributes most of its light upward.

Induced Draft (in a Boiler). Combustion air drawn through the burner or fuel bed by a power driven fan in the flue.

Industrial Equipment. Mechanical or non-mechanical devices used in an industrial setting.

Industrial Fluorescent. A large fluorescent light fixture used in an industrial setting.

Industrial Hygienist. A professional qualified by education, training, and experience to anticipate, recognize, evaluate and develop controls for occupational health hazards.

Industrial Wood Floor. A heavy duty wood floor made of decking (2 inches thick) or of wooden blocks laid on end. Also called a "factory" floor. Very resistant to heavy loads and traffic.

Infiltration. The exchange between conditioned room air and outdoor air through cracks and openings in the building enclosure.

Inlet. An opening for intake.

Insolvency. The financial situation in which liabilities exceed assets.

Instruction to Bidders. More information given to identified bidders than is included in the invitation to bid.

Insulated Block. Hollow masonry block filled with insulation.

Insulated Glass. A type of glass constructed in a manner to protect against sound, heat, heat loss or moisture. Double or triple glazed glass.

Insulation. Any material used to reduce the effects of heat, cold or sound transmission and to reduce fire hazard. Any material used in the prevention of the transfer of electricity, heat, cold, moisture and sound.

Insulation, Rigid. Sheet or board form of rigid foam.

Insulation, Sprayed. A plastic foam or substance sprayed on a surface to insulate.

Insulation, Sill Sealer. Insulation place between sill plate and supporting concrete or masonry to prevent air leaks.

Interceptor. A device that collects foreign matter and prevents it from reaching the sewer system.

Interrogatories. Written questions that a person in a lawsuit or litigation must answer in writing, under oath.

Invert, Manhole. The lowest inside surface of a manhole. A channel in the manhole through which wastewater or stormwater flows.

Inverted Roof. A membrane roof assembly in which the thermal insulation lies above the membrane.

Invitation to Bid. A formal written invitation to submit a bid, usually placed in trade papers or newspapers, informing prospective bidders about a project. Necessary for public work, not necessary for private work.

IPM. Integrated Pest Management.

Irregular Stone. Stone cut to or quarried in different shapes and/or sizes.

Irrigation, Lawn. To supply water to grassy areas by artificial means.

Isolation Joint. A separation between adjoining parts of a concrete structure, usually a vertical plane, at a designed location such as to interfere least with performance of the structure, yet such as to allow relative movement and avoid formation of cracks elsewhere in the concrete and through which all or part of the bonded reinforcement is interrupted.

Isolation System. The collection of structural elements which includes all individual isolator units, all structural elements which transfer force between elements of the isolation system, and all connections to other structural elements. The isolation system also includes the wind restraint system if such a system is used to meet the design requirements.

J

Jack Rafter. A shortened rafter that joins a hip or valley rafter to the top of the wall plate.

Jacking Force. The temporary force exerted by device that introduces tension into prestressing tendons in prestressed concrete.

Jacking Pipe. Forcing pipe through the ground in a tunnel created by the pipe itself. The pipe is generally jacked horizontally in short lengths.

Jamb Anchor. Steel anchor to fasten steel door frame to wall or partition construction.

Jamb. The vertical side of a door or window. The side piece or post of an opening; commonly applied to the door frame.

Jib Crane. A crane which has a projecting arm of its derrick boom.

Jitterbug. A grate tamper for pushing coarse aggregate slightly below the surface of a slab to facilitate finishing.

Job Built Form. A temporary structure or mold constructed on a jobsite, for the support of concrete while it is setting and gaining sufficient strength to be self-supporting.

Job Requirements. A list of specific, necessary and essential tasks to bring to completion a building or structure. Sometimes called "General Requirements".

Joint, Contraction. Formed, sawed, or tooled groove in a structure to create a weakened plane and regulate the location or cracking resulting from the dimensional change of different parts of the structure.

Joint Control. An independent escrow used to safeguard and disburse construction funds.

Joint, Expansion. A joint through tile, mortar, concrete or masonry down to the substrate, intended to allow for gross movement due to thermal stress or material shrinkage. A discontinuity or break intended to allow movement.

Joist. One of a group of light, closely spaced beams used to support a floor deck or flat roof. Timbers supporting the floorboards.

Joist Bridging. The bracing of joists by the fixing of lateral members between them. Pieces fitted in pairs from the bottom of one floor joist to the top of adjacent joists, and crossed to distribute the floor load; sometimes pieces of width equal to the joists and fitted neatly between them. Diagonal or longitudinal members used to keep horizontal members properly spaced, in lateral position, vertically plumb, and to distribute load.

Joist Girder. A light steel truss used to support open-web steel joists.

Joist, Sister. The reinforcement of a joist by nailing, or attaching alongside the existing joist, another joist or reinforcing member.

Joist, Wood. A piece of lumber used horizontally as a support for a floor, measuring two to four inches thick and six or more inches wide.

Journeyman. An experienced reliable worker who has learned his trade and works for another person.

Judgment. A final decision of a court.

Junction Box. A metal box in which runs of cable meet and are protectively enclosed.

Jute Padding. A padding made of a durable yarn that comes from plant fiber used as an underlayment for carpet..

K

K Bracing. That form of bracing where a pair of braces located on one side of a column terminates at a single point within the clear column height.

Keene's Cement. A cement composed of finely ground, anhydrous, calcined gypsum, the set of which is accelerated by the addition of other materials, used in areas subjected to moisture. A hard, strong finishing plaster that is made from gypsum and maintains a high polish.

Kelvin. A scale for registering temperature in which 0 degrees represents absolute zero.

Kerf. The cut made by a saw.

Key. A slot formed into a concrete surface for the purpose of interlocking with a subsequent pour of concrete; a slot at the edge of a precast member into which grout will be poured to lock it to an adjacent member.

Kick Plate. A metal plate or strip that runs along the bottom edge of a door to protect against the marring of the finished surface.

Kiln Dried. Lumber that has been heated in a kiln to dry and control the amounts of moisture.

King Post. In a roof system, the member placed vertically between the center of the horizontal tie beam at the lower end of the rafters and the ridge, or apex of the inclined rafters. A term often used to refer to a certain type of fabricated truss.

Kip. A unit of weight or force equal to 1,000 lbs.

Kitchen Cabinet. A piece of furniture found in the kitchen that contains shelves, drawers and/or doors, and is used for storage.

Kitchen Compactor. A machine in a kitchen that compresses or compacts materials by using hydraulic weight, force or vibration.

Knee Wall. A short wall under the slope of a roof.

Knotty Pine. Pine wood whose knots are exposed, often used for cabinets and interior paneling.

KSI. Kips per square inch.

L

L Cut. A piece of tile cut or shaped to the letter L.

Labor and Material Bond. A bond, secured by the general contractor, which guarantees that the costs for labor and materials for the project will be paid.

Ladder. Frame consisting of two parallel side pieces connected by rungs at suitable distances to form steps on which persons may climb up or down.

Lag Rod. A large diameter rod with a square or hexagonal head.

Laminated Glass. A glazing material consisting of outer layers of glass laminated to, and encasing, an inner layer of transparent plastic.

Laminated Plastic. Sheets of synthetic resin-soaked material bonded by heat and pressure into a stiff board.

Lamp, Incandescent. An electric lamp in which a filament gives off light when heated to incandescence by an electric current.

Landing. A platform part of a staircase system at the top, bottom, or between two flights of stairs.

Landscape. To improve a site by modification of the terrain, the addition of plant materials (grass, trees, etc.), and/or the possible addition of hardscape (walls, patios, paving, etc.).

Lap. The length by which one bar or sheet of fabric reinforcement overlaps another.

Lap Joint. A connection in which two pieces of material are overlapped before fastening.

Latent Defect. A construction defect that is not perceptible by ordinary observation.

Lateral Force. A force acting generally in a horizontal direction, such as wind, earthquake, or soil pressure against a foundation wall.

Lateral Support. A force or structural member that prevents a structure or earthen mass from moving in a lateral or horizontal direction.

Latex. A water based emulsion of a synthetic rubber or plastic obtained by polymerization and used commonly in coatings and adhesives.

Lath. A wood strip, metal mesh, or gypsum board which acts as a backing and/or reinforcing agent for the plaster scratch coat or initial mortar coat.

Lath, Expanded. Metal lath that has been stamped with a pattern and then expanded to create an open grid to accept wet plaster.

Lath, Stucco. Wood strips or metal mesh which form the base for the application of a cement plaster on an exterior wall surface. Sometimes made of gypsum which is manufactured in a board form.

Lathe, Shop. A machine that rotates an object about a horizontal axis which then can be shaped by a fixed cutter or tool.

Lattice. Crossed wood, or steel bars to create a decorative screen..

Lattice Molding. Flat strip molding used in the construction of lattices.

Lauan Veneer. Very thin sheets of wood, in this case Philippine Mahogany.

Lavatory. A basin with drainage and running water primarily used for washing the face and hands. A room with a toilet and wash basin.

Law. A rule of conduct enforced by courts.

Lawn Irrigation. A system where water is transported and distributed to water grass.

Lawsuit. A proceeding in which the jurisdiction of a court is invoked to resolve to a dispute between two or more parties.

Layout Stick. A long strip of wood marked at the appropriate joint intervals for the tile to be used. It is used to check the length, width, or height of the tilework. A common name for this item is idiot stick.

Lb. Symbol for pound or pounds.

Leaching Pit. An excavated hole (pit) that can hold solids but allows liquids to pass through and leach into the ground.

Ledgerboard. The support for the second-floor joists of a balloon-frame house.

Let-In Bracing. Diagonal bracing nailed into notches cut in the face of the studs so as to avoid an increase in the thickness of the wall.

Level. A term describing the position of a line or plane when parallel to the surface of still water; an instrument or tool used in testing for horizontal and vertical surfaces, and in determining differences of elevation.

LH Joist. A type of long-span high strength bar joist.

Liability. Legal responsibility.

Lien. A legal clam by a party against another party for satisfaction of a monetary claim.

Life-Cycle Cost. A cost that takes into account both the first cost and costs of maintenance, replacement, fuel consumed, monetary inflation, and interest over the life of the object being evaluated.

Lift-Slab Construction. A method of building multi-story sitecast concrete buildings by casting all the slabs in a stack on the ground, then lifting them up the columns with jacks and welding them in place.

Lighting. An artificial supply of light or the apparatus providing it.

Lightning Arrester. A device that protects electrical units from lightning and power surges.

Lightweight Aggregate. Aggregate of low specific gravity, such as expanded or sintered clay, shale, slate, diatomaceous shale, perlite, vermiculite, or slag; natural pumice, scoria, volcanic cinders, tuff, and diatomite, sintered fly ash or industrial cinders; used to produce lightweight concrete. Aggregate with a dry, loose weight of 70 pounds per cubic foot or less.

Lightweight Block. A concrete unit constructed of lightweight materials and used to reduce the weight of walls.

Lightweight Concrete. Concrete that achieves a significant reduction in weight by the substitution of lighter materials for the concrete aggregate.

Limestone Panel. A limestone slab, relatively thin with respect to other dimensions, and rectangular in shape.

Limited Partnership. A partnership in which the management authority and liability of some of the partners is limited.

Liner, Flue. Heat-resistant firebrick or other fireclay materials that make up the lining of a chimney.

Liner Panel. A panel used for interior finish.

Link, Beam. That part of a beam in an eccentrically braced frame which is designed to yield in shear and/or bending so that buckling of the bracing members is prevented.

Lintel. The horizontal beam placed over an opening.

Lintel (Cap). A horizontal structural member spanning an opening, and supporting a wall load.

Lintel, Masonry. Masonry member placed within masonry wall or partition to support loads over an opening.

Liquefaction. Transformation of a granular material (soil) from a solid state into a liquid state as a consequence of vibrations induced by an earthquake.

Liquidated Damages. An amount determined by contract in advance of injury to be paid to compensate a party for an injury or damages.

Live Load. Any load that is not permanently applied to a structure. The weight of people, furnishings, machines, and goods in or on a building. The vertical load superimposed by the use and occupancy of a building.

Load. A force that is applied to a body. A weight or force acting on a structure. Applied external force, such as gravity and wind.

Load Bearing. Supporting a superimposed weight or force.

Loader. An excavating machine with a movable bucket or scoop, used to transport earth, crushed stone, or other construction materials.

Lockset. A device installed in a door that has both a deadbolt and doorknob assembly.

Lockwasher. A flat, split ring of metal or steel that when tightened with a nut is used prevent loosening.

Lookout. The end of a rafter, or the construction which projects beyond the sides of a house to support the eaves; also the projecting timbers at the gables which support the rake boards.

Lot Line. The boundary of a parcel of land.

Louver. A construction of numerous sloping, closely spaced slats used to prevent the entry of rainwater into a ventilating opening. A kind of window, generally in peaks of gables and the tops of towers, provided with horizontal slots which exclude rain and snow and allow ventilation.

Low-Lift Grouting. A method of constructing a reinforced masonry wall in which the wall is grouted in increments not higher than 4 feet.

Lugs. Spacers, or protuberances on the sides of ceramic tiles. These devices automatically space the tile for the grout joints.

Lumber. Sawed parts of a log such as boards, planks, and timber. Wood members which are manufactured by sawing, resawing, passing lengthwise through standard planing machine, crosscutting to length, and matching, but without further manufacturing.

Lumen. A quantitative unit for the measurement of the flow of light energy.

Luminaire. A complete lighting unit consisting of a light source, switch, globe, reflector, housing and wiring.

Lump Sum Contract. A contract in which the amount to be paid to the contractor is agreed in advance to be a stipulated sum.

M

Macrozones. Large zones of earthquake activity such as zones designated by the Uniform Building Code Map.

Magnetic Door Holder. A door holder using a magnet to hold it in an open position.

Mahogany Veneer. A thin layer of straight-grained medium density wood for an outer finish or decoration.

Main Breaker. The main electrical service protective device where the power enters a building.

Make-Up Air. Air brought into a building from outdoors through the ventilation system and that has not been previously circulated through the system.

Malice. Desire to do harm; anger, hatred.

Malicious Prosecution. Pursuing a lawsuit without probable cause.

Malleable Iron. Iron that can be hammered or bent without breaking.

Malpractice. Negligent act or omission of a professional.

Manhole. A hole through which a person may go to gain access to an underground or enclosed structure.

Mansard. A roof shape consisting of two superimposed roof lines with the upper level being at a low pitch or almost flat and the lower level at a steep pitch.

Mantle. The decorative outer top or shelf over the inside face of a fireplace. The main bulk of the earth between the crust and core.

Manufactured Roof. A factory-finished roof system.

Manufactured Wall. A factory-finished wall system.

Maple Floor. A floor constructed from a dense, durable wood found from Eastern Canada down to the Southern U.S. Commonly used in gym floors.

Marble. Limestone that is more or less crystallized by metamorphism, that ranges from granular to compact in texture, that is capable of taking a high polish, and that is used in architecture and sculpture.

Masking. The presence of a background noise increased to a level to which a sound signal must be raised in order to be heard or distinguished. When painting, protecting areas not to be painted.

Masonry. Brickwork, blockwork, and stonework.

Mass. The property of an object that causes it to resist changes in its state of motion. This resistance is called inertia.

Masterformat. The copyrighted title of a uniform indexing system for construction specifications, as created by the Construction Specifications Institute and Construction Specifications Canada, commonly (but inaccurately) called the CSI format or numbering system.

Mastic. Organic tile adhesive. A viscous, dough-like, adhesive substance; can be any of a large number of formulations for different purposes such as sealants, adhesives, glazing compounds, or roofing membranes.

Mat, Concrete. A grid of metal reinforcement for concrete foundations, slabs, or mats.

Materialman. An individual or organization who supplies construction materials to a project.

Mat Foundation. A concrete slab used as a building or equipment foundation.

Matching, or Tonguing and Grooving. The method used in cutting the edges of a board to make a tongue on one edge and a groove on the other.

Material Handling. The act or process of transporting materials on or to a jobsite.

Mechanical. Of or relating to machinery or tools. Relating to, governed by, or in accordance the principles of mechanics. Loosely used as a term for anything in the plumbing, heating, air-conditioning, or fire sprinkler trades.

Mechanics Lien. A charge against real estate for the value of work or materials incorporated thereon.

Meeting Rail. The bottom rail of the upper sash of a double-hung window. Sometimes called the checkrail.

Melt. To change a solid into a liquid by the application of heat; or the liquid resulting from such action.

Member. A single piece in a structure, complete in itself.

Membrane. A sheet material that is impervious to water or water vapor.

Membrane Roof. A roof structure with a covering of a sheet material that is impervious to water or water vapor. Commonly a single sheet of material.

Membrane Waterproofing. A membrane, usually made of built-up roofing or sheet material, to provide a positive waterproof floor over the substrate, which is to receive a tile installation using a wire reinforced mortar bed.

Mercury Fixture. A light fixture that has an electric discharge lamp that produces a blue-white light by creating an arc in mercury vapor enclosed in a tube or globe.

Mesh, Slab. Welded-wire fabric in sheets or rolls used to reinforce concrete slabs.

Mesh Tie. A wire used to hold sheets of mesh together so they will not move or spread apart when concrete is poured over the mesh.

Mesh Wire. A series of longitudinal and transverse wires arranged at right angles to each other in sheets or rolls, used to reinforce mortar and concrete. Welded-wire fabric.

Metal. Any of various fusible, ductile and typically lustrous chemical elements that can conduct heat and/or electricity.

Metal Anchor. A bolt or fastener made of metal.

Metal Beam Anchor. Formed steel component used to anchor one end of a beam to another beam, girder, or column and prevent displacement of the beam under lateral or uplift loads.

Metal Building. A building or structure constructed of metal. A term used to denote buildings that are prefabricated in a factory and installed on the job-site.

Metal Chimney. Vertical metal structure with one or more flues to carry smoke and other gases or combustion into atmosphere.

Metal Clad Cable. Electrical conduit of a flexible steel jacket wrapped around insulated wires.

Metal Clad Door. A flush wooden door covered in sheet metal.

Metal Ductwork. The light sheet metal material out of which the ducts of an HVAC system are manufactured.

Metal Fabrication. The building, construction or manufacture of metal structures or metal devices.

Metal Framing. The construction of a building or structure by using steel. Loosely used term to denote the construction of frame houses and partitions by using light gauge metal studs and members.

Metal Halide Lamp. A lamp that uses an electric-discharge to produce light from a metal vapor such as sodium or mercury.

Metal Joist. Horizontal cold formed metal framing member of floor, ceiling or flat roof to transmit loads to bearing points. Often refers to a bar joist.

Metal Lath. A steel mesh used primarily as a base for the application of plaster.

Metal Lintel. A horizontal metal member spanning and carrying the load above an opening.

Metal Pan Stair. A stair assembly constructed to hold precast or cast-in place concrete, masonry, or stone in sheet metal pans at the treads and landings.

Metal Shingle. A roof covering unit manufactured from metal and applied in an overlapping pattern. Metal material used as an exterior wall finish over sheathing.

Metal Stud. Vertical formed steel channel within a framed wall.

Metal Toilet Partition. A prefinished, manufactured dividing wall in a bathroom.

Metal Toothed Ring. Metal rings with toothed edge to embed in wood to resist shear.

Metal Track. Horizontal channel-shaped steel member located at top or bottom to receive metal studs.

Metal Window. Any window manufactured from metal.

Metering. The mechanical process of measuring the usage of water, electricity or gas.

Mild Steel. Steel containing less than three-tenths of one percent carbon, not used as structural steel because of its low strength.

Millwork. Interior components such as trim work, cabinets, doors and windows, etc. but not including floors, siding and ceilings. Often made of wood or plastic laminates and produced in a shop or factory.

Mineral Wool. Any of various lightweight fibrous materials used in heat and sound insulation.

Mirror, Plate. Thick mirror glass manufactured with a high-quality standard.

Misconduct. Wrongful conduct.

Mistake. A legal doctrine under which formation of a contract may be prevented if a party entered into the contract under a material mistake of fact.

Miter. The joint formed by two abutting pieces meeting at an angle.

Mix. The act or process of mixing; also mixture of materials, such as mortar or concrete.

Mixer. A machine used for blending the constituents of concrete, grout, mortar, cement paste, or other mixtures.

Mixer, Plant. An operating installation of equipment including batchers and mixers as required for batching or for batching and mixing concrete materials; also called mixing plant when equipment is included.

Mixing Cycle. The time taken for a complete cycle in a batch mixer, i.e., the time elapsing between successive repetitions of the same operation (e.g., successive discharges of the mixer).

Mixing Speed. Rotation rate of a mixer drum or of the paddles in an open-top, pan, or trough mixer, when mixing a batch; expressed in revolutions per minute (rpm), or in peripheral feet per minute of a point on the circumference at maximum diameter.

Mixture. The assembled, blended, co-mingled ingredients of mortar, concrete, or the like; or the proportions for their assembly.

Mobilization. The act of putting into movement or circulation. The assembly and movement of equipment to a jobsite.

Modification. An agreed change to the terms of a contract.

Modulus of Elasticity. The ratio of the unit stress in a material to the corresponding unit strain. The ratio of normal stress to corresponding strain for tensile or compressive stresses below the proportional limit of the material; referred to as "elastic modulus of elasticity"; "Young's modulus," and "Young's modulus of elasticity"; denoted by the symbol E.

Mogul Base. A screw-in style base for an incandescent lamp of generally 300 watts or more.

Moisture, Absorbed. Moisture that has entered a solid material by absorption and has physical properties not substantially different from ordinary water at the same temperature and pressure.

Moisture Barrier. A membrane used to prevent the migration of liquid water through a floor or wall.

Moisture Protection. The act or process of retarding the seepage of moisture.

Molding. A strip of wood, plastic or plaster with an ornamental profile.

Molding, Lip. A molding with a lip which overlaps the piece against which the back of the molding rests.

Moment. The commonly used expression for the more descriptive term bending moment. A force acting at a distance from a point in a structure so as to cause a tendency of the structure to rotate about that point. The act of a force to cause rotation about a given point or axis.

Moment Connection. A connection between two structural members that is highly resistant to rotation between the members, as differentiated from a pin connection, which allows rotation.

Moment of Inertia. The summation of the products obtained by multiplying each individual unit of area by the square of its distance to an axis.

Monolith. A body of plain or reinforced concrete cast or erected as a single integral mass or structure.

Monolithic Concrete. Concrete cast with no joints other than construction joints or as one piece, generally, the term is used on larger structures.

Mop Sink. A deep well plumbing fixture with a faucet and a drain used for collecting and dispensing water for mopping/janitorial purposes.

Mortar. A mixture of cement paste and fine aggregate; in fresh concrete, the material occupying the interstices among particles of coarse aggregate; in masonry construction, mortar may contain masonry cement, or may contain hydraulic cement with lime (and possibly other admixtures) to afford greater plasticity and workability than are attainable with standard hydraulic cement mortar. A substance used to join masonry units consisting of cementitious materials, fine aggregate, and water.

Mortar Hoe. The mortar hoe is used for hand-mixing mortar. The most common type has a perforated blade and a handle about 66" in length.

Mortar Mixer. A mechanical device for the mixing of mortar. Most mortar mixers are driven by gasoline combustion engines. Electrically driven mixers are used when small batches of mortar are needed.

Mortgage. A lien against real estate that secures payment of a debt.

Mortise. The hole which is to receive a tenon, or any hole cut into or through a piece by a chisel; generally of rectangular shape.

Mortise-and-Tenon. A joint in which a tongue-like protrusion (tenon) on the end of one piece is tightly fitted into a rectangular slot (mortise) in the side of the other piece. A joint made by cutting a hole or mortise, in one piece, and a tenon, or piece to fit the hole, upon the other.

Moss, Peat. Moss containing partially carbonized vegetable tissue formed by the partial decomposition of water in that moss.

Motion. Application to a judge or arbitrator for an order or ruling.

Movable Partition. A dividing wall that can be moved and arranged to form different walled spaces.

MSDS. Material Safety Data Sheet.

Mud. A term used in the trade for mortar.

Mulch. A mixture, as of leaves and compost, that covers or is mixed with the earth, often to help enrich the soil. Bark, crushed stone or other material used to cover planting beds, retain moisture, reduce weeds, and improve appearance.

Mullion. A vertical or horizontal bar between adjacent window or door units. The member between the openings of a window frame to accommodate two or more windows.

Multi-Zone Air Handling System. A system providing conditioned air similar to a single-zone system. The temperature and flow of the air supplying each zone is controlled separately. A constant supply of air is supplied to the various zones.

Muntin. A small vertical or horizontal bar between small lights of glass in a sash. The vertical member between two panels of the same piece of panel work. The vertical sash-bars separating the different panels of glass.

Mutuality. The concept that a contract, to be enforceable at all, must be enforceable by both parties.

N

Nail. A stiff metal wire fastening device with a point on one end and head designed for impact on other end.

Nailable Concrete. Concrete, usually made with suitable lightweight aggregate, with or without the addition of sawdust, into which nails can be driven.

Natural Cleft Slate. Slate which has been split into thinner pieces along its natural cleft or seam. It is rougher in appearance than machined slate.

Natural Convection. Fluid motion of air caused by the temperature difference between the solid surface and the fluid with which it is in contact.

Natural Draft. A natural stream of air up a chimney, caused by the stack effect, which draws combustion air through the burner or fuel bed.

Neat Cement. Hydraulic cement in the unhydrated state.

Neat Cement Grout. A fluid mixture of hydraulic cement and water, with or without admixture; also the hardened equivalent of such mixture.

Neat Cement Paste. A mixture of hydraulic cement and water, both before and after setting and hardening.

NEC. National Electric Code.

Needle Beam. A steel or wood beam threaded through a hole in a bearing wall and used to support the wall and its superimposed loads during underpinning of its foundation.

Negligence. The failure to exercise due care

Neighborhood. In community planning, a residential area in which residents are within walking distance of each other.

NEMA Switch. A electrical switch approved by the National Electrical Manufacturer's Association (NEMA).

Neoprene. A type of synthetic rubber with outstanding oil resistance. Can be used for quick-setting, high strength adhesives.

Neoprene Membrane. An impervious, oil-resistant synthetic rubber, manufactured to be installed in layer form. Used for roofing and waterproofing.

Neoprene Roof. A roof system covered with a sheet of synthetic rubber.

Net, Safety. A meshed fabric that is spread below activity, to protect materials or people that may fall from dangerous heights.

Neutral. A loosely used term for a neutral arbitrator.

Neutral Arbitrator. An arbitrator not controlled by or biased in favor of any party.

Neutral Axis. The line on a member cross-section on which the bending moment is zero.

Newel. The principal post of the foot of a staircase; also the central support of a winding flight of stairs.

Nipple, Bushed. A pipe threaded at both ends to connect two pipes of different dimensions.

Nipple, Offset. A fitting, threaded at both ends, that is a combination of elbows or bends which brings one section of pipe out of line with, but into a line parallel with, another section.

No-Hub Pipe. Pipe usually manufactured of cast iron, which is fabricated without hubs and coupled together by a stainless steel and rubber fastener.

Noise Reduction. The reduction in sound pressure level caused by making some alternation to a sound source. The difference in sound pressure level measured between two adjacent rooms caused by the transmission loss of the intervening wall.

Nominal Size (Lumber). The commercial size designation of width and depth, in standard sawn lumber and glued-laminated lumber grades; larger than the standard actual net size of the finished, dressed lumber.

Non-Metallic Tubing. A round sheath product, of round cross-section, fabricated from a moisture-resistant, flame-retardant material.

Nonaxial. In a direction not parallel to the long axis of a structural member.

Nonbearing. Not carrying a load.

Non-shrink Grout. Cementitious or epoxy based mix used to fill gap created between bearing components or base plates and foundation or other supporting element.

Normal Loading. A design load that stresses a member or fastening to the full allowable stress tabulated. This loading may be applied for approximately 10 years, either continuously or cumulatively, and 90 percent of this load may be applied for the remainder of the life or the member or fastening.

Norman Brick. A brick with 2-3/4 by 4 by 12 inch dimensions.

Nosing. The projecting forward edge of a stair tread. The part of a stair tread which projects over the riser, or any similar projection; a term applied to the rounded edge of a board.

Notched Trowel. A notched trowel with a serrated or square-tooth design. The teeth are made in various sizes. The correct tooth size and depth must be used to apply the thickness of bonding mortar specified. These trowels are used to apply all of the various kinds of bonding materials for ceramic tile.

Notice. Delivery of information to a party

Notice of Completion. A written document, which may be filed at a courthouse in which the project is located, verifying that the project is complete and ready for occupancy. This document initiates the beginning of the last several days of the lien period.

Nuisance. A condition of or on real property that damages neighboring persons or property.

Nut, Hex. A six-sided, short metal nut with a threaded hole for receiving a rod or threaded bolt.

Nylon Carpet. A carpet made from synthetic nylon fibers.

O

Oak. A strong, hard, heavy wood.

Oak Floor. A common type of flooring, usually installed in small tongue and groove slats made of oak.

Oak Veneer. A thin layer of oak glued to an interior wood or bonded together to form a protective or ornamental facing. Types available are white oak and red oak.

Occupancy. A Building Code term referring to the use of a building, such as school, office, residence.

Offer. A promise that is enforceable if accepted.

Office Cubicle. A small compartment unit, surrounded by fixed or movable walls on two or three sides, where an individual works.

Office Partition. Interior walls, which sometimes move, to partition-off an office area, assembled together to create cubicles for employees.

Office Trailer. A highway vehicle parked on a job site, designed to serve wherever needed, as an office and a place to carry out business.

Offset Connector. A fitting that is a combination of elbows or bends which brings one section of pipe out of line with, but into a line parallel with, another section.

Ogee Molding. A molding with a modified S-shaped profile.

Oil Fired Boiler. A boiler that is heated by a unit that sends oil under high pressure to a nozzle, where it is sprayed as a mist and ignited by an electric spark.

Oil Furnace. A furnace that burns heating oil.

One-Way Action. The structural action of a slab that spans between two parallel beams or bearing walls.

Open Type Decking. A deck in which the joists on the underside are exposed.

Open-Web Steel Joist. A prefabricated, welded steel truss used at closely spaced intervals to support floor or roof decking.

Opposed Blade. Two sets of blades in a damper, linked so that the adjacent blades can open and turn in opposite directions.

Oral Agreement. A contract in words that are not reduced to writing.

Ordinance. A law enacted by a city or a county.

Oriented Strand Board (OSB). A building panel composed of long shreds of wood fiber oriented in specific directions and bonded together under pressure. Commonly called "strand board".

Original Contractor. In the law of mechanics liens, a contractor who contracts directly with an owner of real property.

Ornamental Metal. A detail that is added to a building constructed of metal, with the purpose of embellishment or decorating the structure.

OS&Y Valve. A type of valve, with external exposed threads supported by a yoke, indicating the open or closed position of the valve.

Out of Phase. The state wherein a structure in motion is not at the same frequency as the ground motion; or where equipment in a building is at a different frequency from the structure.

Overhang Beam. A beam that is supported by two or more supports and has one or both ends projecting beyond the support.

Overhead. Business costs that cannot be allocated to any specific project operations.

Overhead Door. A door, commonly used in garages and warehouse, that opens upward from the ground.

Owner-Builder. An owner of property who undertakes to construct improvements thereon but does not employ a general contractor.

P

P-Trap. A P-shaped drain connection that prevents sewer gas from escaping from a plumbing fixture.

Packaged Heater. A factory assembled piece of equipment that supplies heat producing units for circulation to a desired location or use.

Packaged Terminal Air Conditioner (PTAC). A self contained unit, operated in the direct expansion method (DX), located in the space served.

Pad, Bearing. A thick slab-type stone or precast concrete block placed under a structural member of a building to spread the load or weight evenly.

Pad, Equipment. Typically, a cast-in-place or precast concrete block placed under mechanical devices to spread the weight and load of the machinery evenly and to prevent excessive vibration.

Pad Eyes. Metal rings mounted vertically on a plate for tying small vessels.

Pad, Transformer. A precast or cast-in-place concrete block placed under a transformer to spread and support its weight.

Pan. A form used to produce a cavity between joists in a one-way concrete joist system.

Panel Cladding. Metal sheathing panels used to provide durability, weathering and corrosion, or impact resistance.

Panel, Limestone. A limestone cladding member, usually machined before installation, relatively thin with respect to other dimensions, and rectangular in shape.

Panel, Siding. Sheet material used as exterior surface or cladding for exterior framed wall to provide protection from exterior element.

Panelboard. A collection of service disconnect switches along with switches to major branch lines (feeders) mounted together in a panel.

Panic Device. A horizontal bar mounted across the full length of a door or a sort of large push-plate, acting as a latching system and operated by inside pressure on the bar or plate. A mechanical device that opens a door automatically if pressure is exerted against the device from the interior of the building. Often called "panic hardware".

Paper Backed Insulation. Insulation that comes with paper facing on one side which serves as a vapor retarder.

Paper Backed Lath. Rectangular sheets of gypsum plaster applied to interior walls, to provide a smooth finished surface with a thin layer of paper acting as a vapor retarder.

Paper, Building. A generic term for felts, papers and sheet materials used in building applications.

Paper, Curing. Waterproof paper placed over freshly finished concrete, to help control the humidity and temperature, aiding in the proper curing of concrete.

Parapet. The region of an exterior wall that projects above the level of the roof.

Parging. Portland cement plaster applied over masonry.

Parking Barrier. A structure either temporary or permanent, that is placed to prevent the encroachment of vehicles.

Parking Stall Painting. The act or process of painting border lines for the parking of individual motor vehicles.

Parol Evidence Rule. The meaning of a contract cannot be changed by reference to statements made by the parties to the contract before the contract was formed.

Parquet Floor. A floor covering laid out in a geometric design composed of small pieces of wood.

Particleboard. A building panel composed of small particles of wood and resins bonded together under pressure. Flat sheet material producing durable and dimensionally stable product which is often used in dry conditions in place of plywood.

Partisan Arbitrator. An arbitrator controlled by, appointed by, or biased in favor of a party to a dispute.

Parting Bead. A narrow vertical strip in a double-hung window frame separating the upper and lower sashes.

Partition. A permanent interior wall which serves to divide a building into rooms.

Partition Toilet. Pre-finished panels used in a toilet enclosure.

Partnership. A business organization in which the owners are mutually responsible for management and liability.

Passenger Elevator. An elevator used for people.

Passive Pressure. The horizontal resistance of the soil to forces against the soil.

Patio Block. Lightweight concrete paving slabs installed in lightly used foot traffic areas.

Pavement Cutting. The process of scoring or cutting through pavement surfaces with a power saw with a specific blade for that purpose.

Pavement Marking. The act or process of applying painted lines or necessary instructional signage on pavement surfaces for pedestrians or vehicle drivers.

Paver, Masonry. Shaped or molded units, composed of stone, ceramic brick or tile, concrete, or cast-in place concrete used for driveways and patios.

Paver, Stone. Blocks of rock processed by shaping, cutting or sizing, used for driveways and patios.

Paving, Concrete. The use of a mixture of portland cement, fine aggregate (sand) coarse aggregate (gravel or crushed stone) and water, to make a hard surface in areas such as walks, roadways, ramps, parking areas, etc..

Payment. Satisfaction, or partial satisfaction, of a debt.

Payment Bond. An guarantee by a surety that those persons who supply work and materials to a construction project will be paid for the work and materials

Payment Schedule. A document, usually attached to a construction contract, that specifies the times for, and amounts of, payments for construction services.

Payroll. The record of wages, salaries, and fringe benefits paid by an employer to its employees.

Pea Gravel. Screened gravel particles most of which would pass through a 3/8 inch (9.5mm) sieve.

Peat Moss. Moss containing a partially carbonized vegetable tissue formed by the partial decomposition in that moss.

Pecan Veneer. A thin layer of hardwood glued over a core of sturdier less valuable solid wood, or plywood, used in flooring and furniture.

Peculiar Risk Doctrine. The doctrine that an owner, by employing an independent contractor, cannot escape liability to persons who may be injured during construction operations on the owner's property, since the construction operations involve a special risk of harm.

Pedestal. A short compression member of reinforced concrete that is placed between a column and the footing to distribute the load to the footing.

Pedestal Floor. A flooring system which has short piers or legs used as a base and the flooring laid over those piers to provide a floor system. Special flooring designed to prevent electrostatic buildup and sparking in a computer room. Usually elevated over the existing floor to facilitate the installation of wires between the components in the room.

Pegboard. A board with holes in which pegs/fittings may be inserted to hang tools or objects.

PELs. Permissible Exposure Limits (standards set by OSHA).

Penalty Clause. A clause in a construction contract by which a contractor is assessed with a monetary penalty, usually on a daily basis, for delay in the completion of a project.

Penthouse Louver. A louvered wall around the mechanical penthouse area of a structure, with fixed or movable flaps. This protects it from the elements and provides a visual screen around equipment.

Perforated Pipe. Pipe with one or more rows of uniform holes along the length. Buried in the ground alongside building foundations or structures, to aid in drainage of groundwater and moisture.

Performance Bond. A bond, secured by the general contractor, which guarantees that the contract will be performed. An undertaking by a surety that a contractor will perform a contract.

Perjury. Lying under oath.

Perlite. Expanded volcanic rock, used as a lightweight aggregate in concrete and plaster, and as an insulating fill.

Personal Property. Property that is not attached to real estate.

Personnel Lift. An elevator for use by persons at a job site, a building, or structure.

Pest Control. The act or process of the placement of devices or spraying of chemicals or powders to control the spread of insects and pests.

Photographs. Used as a loose term for the pictures or photographs taken before a job commences to provide an accurate representation of what the site was like before construction.

Pickup Truck. A light truck having an open body with low sides and tailboard.

Picture Window. A large, often fixed, window, usually of plate or insulating glass.

Pier. Timber, concrete, or masonry supports for girders, posts, or arches. Intermediate supports for a bridge span. Structure extending outward from shore into water used as a dock for ships.

Pilaster. A vertical, integral stiffening rib in a masonry or concrete wall. A portion of a square column, usually set within or against a wall.

Pile. A long, slender, piece of material driven into the ground to act as a foundation. A member embedded into the ground that supports vertical loads, can be made of wood, steel or concrete.

Pile Cap. A thick slab of reinforced concrete poured across the top of a pile group to cause the group to act as a unit in supporting a column.

Pile Spall. A chip or piece broken from a pile by a blow from the driving hammer or by action of the elements.

Pipe. A long tube or hollow body for conducting a liquid, gas, or finely divided solid. May be used for structural elements as well.

Pipe Cleanout. A pipe fitting placed in a drain waste (DWV) system for access to remove blockages.

Pipe Flange. Projecting ring, ridge or collar placed on pipe to strengthen, prevent sliding, or accommodate attachments.

Pipe Insulation. Insulation that covers pipes to help in the reduction of heat loss or gain.

Pipe Jacking. Forcing pipe through the ground in a tunnel created by the pipe itself. The pipe is generally jacked horizontally in short lengths.

Pipe, No-Hub. Pipe manufactured in cast iron, which is fabricated without hubs for coupling.

Pipe Sleeve. Cylindrical insert cast into concrete wall or floor to provide for later passage or anchorage of pipe.

Pipe, Structural. Pipe used in a structure to transfer imposed loads to the ground.

Pitch Board. A board sawed to the exact shape formed by the stair tread, riser, and slope of the stairs and used to lay out the carriage and stringers.

Pitch Pocket. An opening between growth rings of a tree which usually contains resin, bark, or both. In roof construction, a metal container placed around a roof penetration at roof level to receive hot bitumen or caulking and provide a roof seal. Commonly found at columns or plumbing stacks.

Plain Concrete. Concrete that is either unreinforced or contains less reinforcement than the minimum amount specified in the code for reinforced concrete.

Plan. A horizontal geometrical section of a building, showing the walls, doors, windows, stairs, chimneys, columns, etc.. Drawing in which it is assumed that the observer is situated above the area represented.

Plank. A wide piece of sawed timber, usually 1-1/2 to 4-1/2 inches thick and 6 inches or more wide.

Plaster. A cementitious material, usually based on gypsum or portland cement, which is applied to lath or masonry in paste form, to harden into a finished surface. A mixture of lime, hair, and sand, or of lime, cement, and sand, used to cover exterior or interior wall surfaces.

Plaster of Paris. Pure calcined gypsum.

Plaster Ring. A guide with a metal collar attached to a base to apply plaster to a certain thickness or provide a fastener for trim.

Plasterboard. A board used in large sheets as a backing or as a substitute for plaster in walls and consisting of fiberboard, paper, or felt, bonded to a hardened gypsum plaster core. Loose term for drywall or sheetrock.

Plastic Coated Conduit. A type of conduit for electrical wiring that is used around moist areas and highly corrosive fumes.

Plastic Laminate. Sheet material manufactured of multiple layers of paper with top layer of plastic usually 1/16 inch (1.59 mm) thick with decorative finish. Plastic laminate may be used in flat sheets or heat formed bent, and adhered to single curved base material. Commonly referred to by the brand name of "Formica".

Plastic Pipe. Pipe manufactured from hard plastic to resist corrosion and rust.

Plasticity. Property of a material to deform under load and to retain the deformation after the load is removed.

Plate. The top horizontal piece of the wall of a frame building upon which the roof or other structural elements rests.

Plate Cut. The cut at the bottom end of a rafter to allow it to fit upon the plate. The cut in a rafter which rests upon the plate; sometimes called the seat cut.

Plate Girder. A large beam made up of steel plates, sometimes in combination with steel angles, welded, bolted or riveted together.

Plate Glass. Glass of high optical quality produced by grinding and polishing both faces of a glass sheet.

Plate, Toe. A metal bar fastened to the outer edge of a grating; rear of a tread; the bottom rail of a door.

Platform. Horizontal landing in stair either at the end of a flight or between flights, either at floor level or between floors.

Platform Frame. A wooden building frame composed of closely spaced members nominally 2 inches in thickness, in which the wall members do not run past the floor framing members. Typical wood stud wall framing in which the studs are one level in height and the floor framing above rests on the top plates of the wall below. The most common type of framing used in house construction.

Pleadings. Formal written documents, filed with a court, accusing a party of wrongdoing, or defending a party against such an accusation.

Plenum. Air compartment connected to a duct or ducts.

Plexiglass. A plastic resilient material comparable to glass in use, usually manufactured in sheets.

Plow. To cut a groove running in the same direction as the grain of the wood.

Plumb. Vertical, or perfectly straight up-and-down.

Plumb Cut. Any cut made in a vertical plane; the vertical cut at the top end of a rafter.

Plumbing. The act or process of installing in a building or structure the pipes, fixtures, or other apparatus for supplying potable water and removing liquid and water-borne wastes. The installed fixtures and piping of a building or structure.

Plumbing Fixture. Plumbing equipment, usually installed last, such as sinks, water closets, shower/bath units, etc..

Ply. A term used to denote a layer or thickness of building or roofing paper as two-ply, three-ply, etc.

Plywood. A wood panel composed of a number of layers of wood veneer bonded and glued together under pressure. Flat sheet material built up of sheets of veneer called plies, united under pressure by bonding agent to create a panel with adhesive bond between plies as strong as or stronger than the wood alone.

Plywood, Finish. The finest grade of plywood.

Plywood Sheathing. A flat panel made up of a number of thin sheets, or veneers, of wood used to close up side walls, or roofs preparatory to installation of finish materials on the surface.

Plywood Shelving. Horizontal mounted plywood surfaces upon which objects may be stored, supported, or displayed.

PM. Preventive Maintenance.

Pneumatic Concrete. Concrete that is delivered by equipment powered by compressed air.

Pneumatic Tool. A tool powered by compressed air.

Pocket, Pitch. An opening between growth rings of a tree which usually contains resin, bark, or both. In roof construction, a metal container placed around a roof penetration at roof level to receive hot bitumen or caulking and provide a roof seal. Commonly found at columns or plumbing stacks.

Pointing. The process of applying and compacting mortar to the surface of a mortar joint after the masonry has been laid, either as a means of finishing the joint or to repair a defective joint.

Pole. One of the two terminals of an electric cell, battery or dynamo. Either extremity of the axis of a sphere. The vertex of an angle coordinate. A point of guidance. A stake.

Pole, Utility. A vertical pole usually with cross arms, where utilities install their service or supply lines.

Polyethylene Pipe. Pipe manufactured from a thermoplastic, high-molecular-weight organic compound.

Polyethylene Vapor Barrier. A sheet form thermoplastic membrane, high molecular weight, organic compound, used as a protective cover to prevent the passage of air and/or moisture.

Polypropylene Pipe. A tough plastic pipe with resistance to chemicals and heat.

Pop-Up Head. The part of a sink drain assembly that is operated by a linkage to open or close the drain. In lawn irrigation systems, a watering head that retracts when not in use, flush with the ground.

Porous Fill. Soil that allows relatively free passage of water.

Portal Frame. A rigid frame; two columns and a beam attached with moment connections.

Portland Cement. The element used as the binder in concrete, mortar, and stucco.

Post. A timber set on end to support a wall, girder, or other member of the structure.

Post Hole. The dug out holes in the ground for the installation of fence and gate posts.

Post Line. The line which marks the outside face of a foundation, the location line of fence posts or the line of piers for a deck.

Posts, Treated. Vertical posts infused or coated with sealers or chemicals to retard fire, decay, insect damage or deterioration due to weather.

Post-Tensioning. A method by which concrete is compressed after it has been cast by stressing the steel reinforcing. The compressing of the concrete in a structural member by means of tensioning high-strength steel tendons against it after the concrete has cured.

Pour. To place concrete; a continuous increment of concrete casting carried out without interruption.

Power Tool. An apparatus or device used in construction, powered by electric current.

Precast. A concrete component or member cast and cured in other than its final position.

Precast Specialty. Special shapes or ornamental objects of precast masonry or concrete. A loose term which may refer to any manufacturing of precast members.

Prefinished Gypsum Board. Gypsum board finished at the factory with a decorative layer of paint, paper, or plastic.

Prehung Door. A door that is hinged to its frame in a factory or shop.

Preservative Treated. Applied or pressurized chemical treatment of wood or plywood to make it resistant to deterioration from moisture and insects.

Pressure Reducing Valve. A valve which maintains fluid pressure uniformly on its outlet side as long as pressure on the inlet side is at or above a design pressure.

Pressure-Treated Lumber. Lumber that has been impregnated with chemicals under pressure, for the purpose of retarding either decay or fire.

Prestressed Concrete. Concrete that has the reinforcing pretensioned prior to placement.

Prestressing. Applying compressive stress to a concrete structural member to increase its strength.

Pretensioning. A method by which the design tensile force is applied to the steel reinforcing before the concrete is set.

Prevailing Wage Law. A law that establishes minimum wages for job classifications in the construction trades.

Prime Contractor. A general contractor who contracts with a property owner and, in turn, employs a subcontractor or subcontractors to perform some or all of the work.

Processor, Photo. A piece of equipment used to develop photographs from negatives.

Professional Fee. The amount of money charged by a person hired to perform a service.

Profit. The excess of returns over expenditures in transactions or series of transactions.

Projecting Sash. A window with sashes that swing either inward or outward.

Proposal. The document submitted by the contractor to the owner for construction of the project, sometimes called "the bid".

Pro Rata. Proportionate or pro-rated.

Protected Membrane Roof. A membrane roof assembly in which the thermal insulation lies above the waterproof membrane.

Protective Board. A board or sheet of material that is installed next to a waterproofing membrane and then backfilled against thus protecting the membrane from puncture or damage.

Provisional Remedies. Orders or proceedings that protect persons or property while legal proceedings are pending.

PSF. Pounds per square foot.

PSI. Pounds per square inch.

Public Utility Regulatory Policies Act. A law requiring public utility companies to buy on-site generated electricity from private power producers.

Publication. The process of making writings or drawings available for distribution to or inspection by the public.

PUD. A cluster development that includes more land uses in the built up area (industrial, commercial) than the residential cluster system.

Pull Box. A box, in electrical work, placed in a length of conduit, through which cables can be pulled.

Pump. A device that raises, transfers, or compresses fluids or attenuates gases by suction or pressure or both.

Pump, Sump. A small capacity pump that empties pits receiving water, sewage, or liquid waste.

Pumped Concrete. Concrete that is pumped through a hose or pipe.

Punitive Damages. Damages awarded to a private person against a wrongdoer by way of punishment, and to deter future misconduct.

Purger, Air. A mechanical device that removes unwanted air.

Purlin. A timber supporting several rafters at one or more points. Beams or struts that span across a roof to support the roof framing system.

Push Plate. A metal plate on a door installed at hand level to serve as an opening device, used for protection from damage and easier cleaning.

PVC. Polyvinyl Chloride.

PVC Waterstop. A nonmetallic synthetic resin prepared by polymerization of vinyl chloride inserted across a joint to obstruct the seepage of water through the joint.

Q

Quadruplex Cable. Four wire cable.

Quarry Tile. A large clay floor tile, usually unglazed.

Quarter Round. Wood molding in the shape of a quarter circle.

Quasi Contract. Application of contract rules of law to parties that do not have a contractual relationship.

R

R. Represents the radius of gyration.

R-Value. The number of minutes (seconds) required for 1 Btu (joule) to penetrate one square foot (square meter) of a material for each degree of temperature difference between the two sides of the material. The resistance of a material to the passage of heat. The reciprocal of conductors (1/c).

Rabbet. A corner cut out along edge of a piece of wood.

Raceway. Any channel courses supporting and protecting electrical conductors, including conduits, wireways, surface metal raceway, cable trays, floor and ceiling raceways, busways and cable bus.

Radiant Heat. Heat transmitted by radiation.

Radius of Gyration. A characteristic of the cross-section of the member used in the determination of its structural characteristics.

Raft. A footing or foundation, usually a large thick concrete mat.

Rafter. A framing member that runs up and down the slope of a pitched roof. The beams that slope from the ridge of a roof to the eaves and make up the main body of the roof's framework.

Rafter Anchor. A bolt or fastening device which attaches the parallel beams used to support a roof covering to the rafter plate.

Rafter, Hip. The diagonal outside intersecting rafter at the meeting planes in a hip roof. A structural member of a roof forming the junction of an external roof angle or, where the planes of a hip roof meet.

Rafter, Jack. A rafter that fills the space between the hip rafter and the top of the wall plate.

Rafter, Valley. The rafter extending from an inside angle of the plates toward the ridge or center line of the house.

Rafters, Common. Those which run square with the plate and extend to the ridge.

Rail. The horizontal members of the balustrade or panel work.

Railing. An open fence or guard for safety, made of rails and posts. A banding in cabinetwork. On plywood, the solid wood band around one or more edges.

Railing, Pipe. A metal railing made of pipe.

Rake. The sloping edge of a pitched roof. The trim of a building extending in an oblique line, as rake dado or molding.

Ranch Molding. An architectural style of molding that is installed as finish work or ornamentation. A type or style of trim which is gently curved and devoid of ornate design.

Random Width Flooring. Flooring materials of wood with varying widths, often called a "plank floor".

Reaction. In structures, the response of the structure to the loads. The reactions usually refers to the components of force developed at the supports. The force that acts at a support of a structure.

Real Estate, Real Property. Land that is subject to ownership, with its permanent improvements and appurtenances.

Reasonable Use. A legal doctrine that allocates water rights.

Receiver. An officer of the court who administers property pending resolution of dispute between claimants to the property.

Receptacle, Cable. An interrupting outlet box device installed in an electric supply circuit for the connection of electric cables.

Receptor. A metallic or nonmetallic waterproof base for a shower stall.

Recessed Light Fixture. A light fixture installed in a suspended ceiling or a recess in a plaster or gypsum board ceiling.

Record Drawings. A set of drawings prepared by the general contractor, which includes any revisions in the working drawings and specifications during construction, indicating how the project was actually constructed.

Reducer. A pipe fitting which connects pipes of different sizes. A tile trim unit used to reduce the radius of a bullnose or a cove to another radius or to a square.

Redundant Member. Any member of a truss not required for truss stability.

Redwood Siding. Wood species that is resistant to decay used for the final covering of an exterior wall surface of a building or structure.

Reflective Coated Glass. Glass onto which a thin layer of metal or metal oxide has been deposited to reflect light and/or heat.

Reflectorized Sign. A sign made with highly reflective material.

Refractories. Heat-resistant non-metallic ceramic materials.

Refrigeration. The cooling of a material or space.

Refuse Driven Fuel (RDF). A method of recovering heat, in an incinerator, from solid waste.

Register. An opening to a room or space with a grille and a damper at the end of ductwork for the distribution of conditioned air. May be mounted in a ceiling, wall, or floor.

Reglet, Roof. A thin narrow groove cut or cast into a building to accept pieces of flashing.

Regulations. Laws that are enacted by a public administrative agency, rather than by an elected legislative body.

Reinforced Brick Masonry (RBM). Brickwork into which steel bars have been embedded to impart tensile strength to the construction.

Reinforced Concrete. Concrete containing adequate reinforcement (pre-stressed or not prestressed) and designed on the assumption that concrete and steel act together in resisting forces. Concrete work into which steel bars have been embedded to impart tensile strength to the member.

Reinforced Grouted Masonry. Wall construction consisting of brick or block that is grouted solid throughout its entire height and has both vertical and horizontal reinforcing.

Reinforcement. The action or state of strengthening by additional assistance, material, or support. The action or state of strengthening or increasing by fresh additions. A term used for reinforcing steel.

Reinforcement, Mesh. A series of longitudinal and transverse wires arranged at right angles to each other in sheets or rolls, used to reinforce mortar and concrete. Welded-wire fabric.

Reinforcing Accessory. The items used to install reinforcing in concrete. These include but are not limited to; chairs, couplings, tie wire etc.

Release Agent. Material used to prevent bonding of concrete to a form surface.

Relief. Damages or court orders awarded in litigation.

Reline Pipe. To install new linings in pipes. Commonly includes the cleaning of built-up scale or debris from the existing pipe and relining with a compatible material.

Remove. The act or process of demolishing, dismantling, and carrying away an old fitting or component.

Rental Value. A measure of damages for deprivation of the use of property.

Res Judicata. The doctrine that courts will not re-litigate the same issues between the same parties.

Rescission. A legal remedy for a material breach of contract under which the law pretends that a contract never existed and attempts to put the parties in the positions they occupied before the contract was executed, a cancellation.

Resolving Forces. Replacing a force or forces with two or more other forces that yield the same effect on a structure as the original forces.

Restoration. The act or process of bringing a structure back to its former position or condition. The act of restoring to an original condition.

Resultant. A force that will produce the same effect as two or more combined forces.

Retaining Wall. A wall that is designed to resist the lateral pressures of retained soil. A wall that holds back a hillside or is backfilled to create a level surface.

Retardant, Fire. A material or treatment which effects a reduction in flammability and in spread of fire.

Return. The ending of a small splash wall or a wainscot at right angle to the major wall. The continuation of a molding or finish of any kind in a different direction. In HVAC, a term for the return-air duct of a forced air heating/cooling system.

Revolving Door. Typically a four panel door attached at 90 degrees to each other that turns on a center axis.

Ribband. The support for the second-floor joists of a balloon-frame house. The horizontal member of a wood frame wall which supports floor joists or roof rafters.

Ridge. The top edge or corner formed by the intersection of two roof surfaces. The board against which the tips of rafters are fastened.

Ridge Cut. Any cut made in a vertical plane; the vertical cut at the top end of a rafter.

Ridge Vent. A construction element mounted along the ridge of a roof to aid in ventilating an attic space.

Rigid Frame. Two columns and a beam or beams attached with moment connections; a moment-resisting building frame. The connections between the beam and columns are constructed so as to transmit moments.

Rigid Insulation. Thermal insulating material made of polystyrene, polyurethane, polyisocyanurate, cellular glass, or glass fiber, sometimes offered with a skin surfacing, formed to flat board shape of constant thickness.

Rigidity. Relative stiffness of a structure or element.

Riprap. A foundation or sustaining wall of stones placed together without order, as in deep water or on an embankment, to prevent erosion.

Rise. The vertical distance through which anything rises, as the rise of a roof or stair.

Riser. A single vertical increment of a stair; the vertical face between two treads in a stair; a vertical run of plumbing, wiring or ductwork.

Rivet. A structural fastener on which a second head is formed after the fastener is in place.

Rock Anchor. A post-tensioned rod or cable inserted into a rock formation for the purpose of anchoring a structure to it.

Rock Drilling. The act or process of boring holes into rock.

Rodding. A method of using a straightedge to align mortar with the float strips or screeds. This technique also is called dragging, pulling, floating, or rodding off.

Roll Roof. A roof that has been covered with an asphaltic material that comes in rolls.

Roll-Up Door. A door which raises and rolls into a coiled configuration and lowers on tracks on either side.

Roof. The top cover of a building or structure.

Roof Ballast. Crushed rock or gravel which is spread on a roof surface to form its final surface.

Roof Bond. A legal guarantee that a roof installed is in accordance with specifications and will be repaired or replaced if it fails within a certain period of time due to normal weathering.

Roof, Built-Up. A roof covering made of continuous rolls or sheets of saturated or coated felt embedded in a bituminous coating. The roofing may have a final ballast coating of gravel or slag.

Roof Cricket. Relatively small elevated area of wood roof constructed to divert water around chimney, curb, or other projection.

Roof, Curb. A roof with a slope that is divided into two pitches of each side. Also known as a gambrel or mansard roof.

Roof, Fluid-Applied. A roof coated with an asphalt-based liquid.

Roof Insulation. Materials used in between rafters or roof supports for the protection from heat and cold. Solid sheets of insulating material installed on a flat roof.

Roof Scupper. A fitting cut through a roof system, acting as a waterway so water can be discharged off the roof.

Roof Sheathing. The first layer of covering on a roof, fastened to the rafter boards, used to support the roofing material.

Roof Shingle. Roof covering pieces of asphalt, asbestos, wood, tile, slate, etc. Pieces of wedge shaped wood or other material used in overlapping courses to cover a roof.

Roof Slab. The flat section of a reinforced concrete roof, supported by beams, columns, or other framework.

Roof Specialties. A fitting or piece of trim used in the installation of a roof, such as gravel stop, flashing, vent strips, etc..

Roof Walkway. A permanent aisle for safe access across a roof. Also serves as a protection for the roofing material when maintenance is being done.

Roofing. The material put on a roof to make it weatherproof.

Rooftop AC. Air-conditioning system or equipment installed or placed on a roof.

Rooftop Heater. A heating unit installed or placed on a roof.

Rooftop Unit. A system of conditioning air in a self contained unit, installed directly above the conditioned space, in a weatherproof enclosure.

Rosewood Veneer. A thin layer of cabinet wood with a dark red or purplish color streaked and variegated with black, applied to give a superior and decorative surface.

Rough Opening. The clear dimensions of the framed opening that must be provided in a wall to accept a given door or window unit.

Roughing In. The act of preparing a surface by applying tar paper and metal lath (or wire mesh). Sometimes called "wiring". In plumbing and electrical, the pre-wiring or pre-piping of a building before the finished walls, fixtures, and devices are installed.

Round, Quarter. Molding in the shape of a quarter circle.

Rowlock. A brick laid on its long edge, with its end exposed in the face of the wall.

Rubble Masonry. Uncut stone, used for rough work, foundations, backfilling, and the like.

Run. Horizontal dimension in a stair. The length of the horizontal projection of a piece such as a rafter when in position.

Runner Channel. A steel member from which furring channels and lath are supported in a suspended plaster ceiling.

Running Bond. Brickwork consisting entirely of stretchers.

Rusticated Concrete. Beveled edges of concrete making the joints conspicuous.

S

Saddle Board. The finish of the ridge of a pitch-roof house. Sometimes called comb board.

Safety Chain. Chain installed horizontally in railing assembly to provide for ease in providing temporary opening in railing.

Safety Glass. Specific type of glass having the ability to withstand breaking into large jagged pieces.

Safety Net. A woven meshed fabric that is spread below activity to protect materials and people that may fall from dangerous heights.

Safety Nosing. Stair nosing with abrasive non-slip strip surface flush with tread surface.

Salamander. A portable source of heat, customarily kerosene or oil-burning, used to temporarily heat an enclosure. Commonly used around newly placed concrete to prevent freezing.

Sales Tax. A tax levied on sales of goods and services which is calculated as a percentage of the purchase price and collected by the seller.

Sample Soil. A representative specimen of soil from a site.

Samples. Material samples requested by the architect of the general contractor or materials specified for the project.

Sampling. The method of obtaining small amounts of material for testing from an agreed-upon lot.

Sandblast. A system of cutting or abrading a surface such as concrete by a stream of sand ejected from a nozzle at high speed. Compressed air is used to propel a stream of wet or dry sand onto the surface. Often used for cleanup of horizontal construction joints or for exposure of aggregate in architectural concrete. A method of scarifying the surface of concrete or masonry to provide a bondable surface. Also used to clean metal before painting.

Sandstone. A sedimentary rock formed from sand.

Sanitary Piping. Drain, waste, and vent plumbing systems.

Sash. A frame that holds glass. The framework which holds the glass in a window.

Sash Bead. A strip with one edge molded, against which a sash slides.

Saw, Band. A saw in the form of a steel belt running over pulleys.

Saw Cut, Concrete. A cut in hardened concrete utilizing diamond or silicone-carbide blades or discs.

Scab. A short piece of lumber used to splice, or to prevent movement of two other pieces.

Scaffold or Staging. A temporary structure or platform enabling workmen to reach high places. A temporary structure for the support of deck forms, cartways, or workmen, or a combination of these such as an elevated platform for supporting workmen, tools, and materials; adjustable metal scaffolding is frequently adapted for shoring in concrete formwork.

Scale. A short measurement used as a proportionate part of a larger, actual dimension. The scale of a drawing is expressed as 1/4 inch = 1 foot.

Scarifier. A piece of thin metal with teeth or serrations cut in the edge. It is used to roughen fresh mortar surfaces to achieve a good bond for the tile. A scarifier also can be used to roughen the surface of concrete.

Scheduling Software. Computer software used to itemize the sequence of project tasks.

Scotia. A hollow molding used as a part of a cornice, and often under the nosing of a stair tread.

Scratch Coat. The first coat in a three-coat application of plaster.

Scratcher. Any serrated or sharply tined object that is used to roughen the surface of one coat of mortar to provide a mechanical key or bond for the next coat.

Screed. A strip of wood, metal, or plaster that establishes the level to which concrete or plaster will be placed. To strike or plane off wet mortar or concrete which is above the desired plane or shape.

Scribing. The marking of a piece of wood to provide for the fitting of one of its surfaces to the irregular surface of another.

Scupper. An opening through which water can drain from the edge of a flat roof. An opening cut through a roof system, acting as a waterway so water can be discharged off the roof.

Sealant. An elastomeric material that is used to fill and seal cracks and joints. At expansion joints, this material prevents the passage of moisture and allows horizontal and lateral movement.

Seat Cut. The cut at the bottom end of a rafter to allow it to fit upon the plate.

Seated Connection. A connection in which a steel beam rests on top of a steel angle fastened to a column or girder.

Section. A drawing showing the kind, arrangement, and proportions of the various parts of a structure. It is assumed that the structure is cut by a plane, and the section is the view gained by looking in one direction.

Security Grille. Steel grille to prevent intrusion or entry through openings.

Seismic. Relating to earthquakes.

Seismic Load. A load on a structure caused by movement of the earth relative to the structure during an earthquake.

Self Furring Lath. Metal lath with dimples that space the lath away from the sheathing behind it to allow plaster to penetrate the lath.

Self Tapping. Creates its own screw threads on the inside of a hole.

Septic Tank. A private sewage system holding tank, installed together with a leaching field, that collects sewage and allows the solid waste to settle to the bottom of an area while the liquid particles of the sewage drain into the leaching field area.

Service Disconnect (or, Main Switch). A switch for, or a means of disconnecting an entire building from electrical service.

Service Sink. A deep basin with a faucet used in janitorial applications.

Service Wye. A pipe fitting which joins three pipes at a 45 degree angle. A drainage fitting in a plumbing system.

Set Screw. A headless screw used to secure two separate parts in a relative position to one another, preventing the independent motion of either part. A screw to adjust the tension of a spring.

Setoff. An offset against a claim.

Setting Bed. The layer of mortar on which the tile is set. The final coat of mortar on a wall or ceiling also may be called a setting bed.

Setting Time. The time required for a freshly mixed cement paste, mortar or concrete to achieve initial or final set.

Shaft. An unbroken vertical passage through a multistory building, used for elevators, wiring, plumbing, ductwork, etc.

Shaft Wall. A wall surrounding a shaft, also used loosely as a term for a certain type of drywall system (firewall) which is used to enclose a shaft.

Shake Roof. A roof covering system made up of thicker, hand-cut cedar wood shingles (shakes).

Shear. A force effect that is lateral, perpendicular to the axis of a structure.

Shear Panel. A vertical plane (wall) that resists lateral forces.

Shear Plate. In heavy timber construction, a round steel plate used for connecting wood to non-wood materials.

Shear Strength. The strength of an element to resist shear.

Shear Wall. A wall designed to resist lateral forces parallel to the wall. A wall portion of a structural frame intended to resist lateral forces, such as earthquake, wind, and blast, acting in or parallel to the plane of the wall.

Sheathing. The rough covering applied to the outside of the roof, wall, or floor framing of a structure. In clay tile or wood shingle/shake roofs, roofing boards generally installed as narrow boards laid with a space between them, according to the length of a shingle or tile exposed to weather.

Shed. A building or dormer with a single pitched roof.

Sheet Floor. A term denoting any type of resilient flooring material that is manufactured and installed in sheets or rolls.

Sheet Glass. Window glass.

Sheet Metal. Flat rolled metal less than 1/4 inches in thickness.

Sheet Metal Roof. A roof covering of aluminum, copper, stainless steel, or galvanized metal sheets. In high-finish applications, loosely referred to as an "architectural roof".

Sheet Piling. Planking or sheeting made of concrete, timber, or steel that is driven in, interlocked or tongue and grooved together to provide a tight wall to resist the lateral pressure of water, adjacent earth or other materials.

Sheetrock. Plasterboard sheets, an interior facing panel consisting of a gypsum core sandwiched between paper faces, also called gypsum board or plasterboard. Different types are available for standard, fire-resistant, water-resistant, and other applications.

Sheetrock Finishing. The final sanding and coating of sheetrock (gypsum board) seams to make ready for painting.

Shelf. A horizontal mounted surface upon which objects may be stored, supported, or displayed.

Shelf Life. Maximum interval during which a material may be stored and remain in a usable condition.

Shell (Concrete). Concrete outside the transverse reinforcement confining the concrete.

Shield, Expansion. Device inserted in predrilled holes (usually in concrete or masonry) which expands as screw or bolt is tightened within it, used to fasten items to concrete or masonry.

Shim. A thin piece of material placed between two components of a building to adjust their relative positions as they are assembled; to insert shims.

Shingle. A small thin piece of building material often with one end thicker than the other for laying in overlapping rows as a covering for the roof or sides of a building or structure.

Shiplap. A board with edges rabbeted so as to overlap flush from one board to the next.

Shop Drawings. Detailed fabrication and/or construction drawings of specific items of a project provided by subcontractors.

Shop Painted. Coating(s) of paint applied in shop, usually a primer coat to protect metal from corrosion, which may or may not receive additional coats in the field.

Short Circuiting. In electrical, an interrupted circuit. In HVAC, a situation that occurs when the supply air flows to exhaust registers before entering the breathing zone. To avoid short-circuiting, the supply air must be delivered at a temperature and velocity that results in mixing throughout the space.

Shotcrete. A low-slump concrete mixture deposited by being blown from a nozzle at high speed with a stream of compressed air.

Shower Compartment. An enclosure in which water is showered on a person.

Shower Pan. A prefabricated assembly to provide a bottom for a shower, may be made of sheet metal, plastic, or masonry. Terminology used in some areas for a waterproof membrane.

Shower Receptor. The floor and side walls of the shower up to and including the curb of the shower.

Shrinkage. The decrease in volume, or contraction, of a material by the escape of any volatile substance, or by a chemical or physical change in the material.

Shrinkage Cracking. Cracking of a structure or member due to failure in tension caused by external or internal restraints as reduction in moisture content develops.

Sidelight. A tall, narrow window alongside a door.

Sidewalk. A walk for pedestrians at the side of a street.

Siding. The outside finish of an exterior wall.

Signage. Any of a group of posted commands, warnings or directions.

Sills. The horizontal timbers of a house which either rest upon the masonry foundations or, in the absence of such, form the foundation.

Simple Beam. A beam that is supported on two supports and where no bending is transferred from the beam to the support.

Sink. A basin with a drainage system and water supply, used for washing and drainage. A pit or pool used for the deposit of waste.

Sister Joist. The reinforcement of a joist by nailing, or attaching alongside the existing joist, another joist or reinforcing member.

Sitecast Concrete. Concrete that is cast-in-place.

Sizing. Working material to the desired size; a coating of glue, shellac, or other substance applied to a surface to prepare it for painting or other method of finish.

Skim Coat. A thin coat of plaster over any base system. May be the final or finish coat on plaster base or a certain type of drywall.

Slab Bolster. Continuous, individual support used to hold reinforcing bars in the proper position.

Slab Form. The formwork used for the pouring or placing of a concrete slab. A type of manufactured metal decking which is made expressly to receive a final layer of poured concrete.

Slate. A form of geologically hardened clay, easily split into thin sheets.

Sleeper. A timber laid on the ground to support a floor joist. A framing system, usually for a wood floor system, which is fastened directly to a concrete floor thus facilitating the installation of the finished floor.

Sliding Glass Door. An exterior glass door mounted above and below on tracks for ease in movement.

Slip Coating. A ceramic material or mixture other than a glaze, applied to a ceramic body and fired to the maturity required to develop specified characteristics.

Slot. An opening in a member to receive a connection with another part.

Slot, Anchor. A groove in an object into which a fastener or connector is inserted to attach objects together.

Slump. A measure of consistency of freshly mixed concrete, mortar, or stucco.

Slump Cone. A mold in the form of the lateral surface of the frustum of a cone with a base diameter of 8 in. (203 mm), top diameter 4 in. (102 mm), and height 12 in. (305 mm), used to hold a specimen of freshly mixed concrete for the slump test; a cone 6 in. (152 mm) high is used for tests of freshly mixed mortar and stucco.

Slump Test. The procedure for measuring slump with a slump cone.

Slurry. A mixture of water and any finely divided insoluble material, such as portland cement, slag, or clay in suspension. A watery mixture of insoluble materials.

Smelt. To melt or fuse ore and to separate metal.

Smelter. A furnace in which the raw materials are melted.

Smoke Detector. A device that detects the presence of smoke, usually resulting in the sounding of an alarm.

Sodium Fixture. An electric lamp that contains sodium vapor and electrodes between which a luminous discharge takes place, commonly used outdoors.

Soffit. The underside part of a member of a structure, such as a beam, stairway, roof, or arch. The undersurface of a horizontal element of a building, especially the underside of a stair or a roof overhang.

Soil. A generic term for unconsolidated natural surface material above bedrock.

Soil and Waste Pipe. Plastic, copper, cast iron or DWV drainage, water and vent.

Soil Gases. Gases that enter a building from the surrounding ground (e.g., radon, volatile organics, pesticides).

Soil Report. A geological report from a geological engineer providing information on the subsurface soil conditions.

Soldier. A brick laid on its end, with its narrow face toward the outside of the wall.

Soldier Course. Oblong tile or brick laid with the long side vertical and all joints in alignment.

Sole Plate. The horizontal piece of dimension lumber to which the bottom of the studs are attached in a wall of a light frame building.

Solid Block. A block with small or no internal cavities.

Solvent. In a solution, that substance which dissolves another is called the solvent. Solvent is also a common term for many liquids which are commonly used in making solutions, e.g., organic solvents, petroleum solvents, etc. Also used for thinning down a fluid, and for cleaning purposes.

Sound Absorption. The process of dissipating or removing sound energy. The property possessed by materials, objects, and structures such as rooms, of absorbing sound energy.

Sound Attenuation. A process in which sound is reduced as its energy is converted to motion or heat.

Space Frame, Space Truss. A truss that spans with two-way action.

Spall. A fragment, usually in the shape of a flake, detached from a larger mass by a blow, by the action of weather, by pressure, or by expansion within the larger mass.

Span. The distance between supports for a beam, girder, truss, vault, arch or other horizontal structural device; to carry a load between supports. The distance between the bearings of a timber or arch.

Spandrel. The wall area between the head of a window on one story and the sill of a window on the floor above; the area of a wall between adjacent arches.

Spandrel Beam. A beam that runs along the outside edge of a floor or roof.

Special Conditions. A written document that revises or clarifies the general conditions, and/or indicates or clarifies project specific conditions. Contains specific conditions of the site or contract.

Special Door. A door that has a unique use, such as a bank vault door.

Specialties. A designation of construction materials or components which commonly typifies items that furnish and finish off the structure. A loosely used term to denote any item which aids in the installation of a structure's parts (i.e. roof specialties).

Specialty Contractor. A contractor who follows a recognized trade: trade contractor or subcontractor. A specialty contractor commonly installs certain specific items such as flooring, windows, terrazzo, etc..

Specifications. A description of systems, materials, and facilities to be employed in building a structure, and incorporated into a contract for construction. A written document for the project indicating quality and supplementing the working drawings. The written or printed directions regarding the details of a building or other construction.

Spiral Reinforcing. Continuously wound reinforcing in the form of a cylindrical helix.

Splash Block. A small precast block of concrete or plastic used to divert water at the bottom of a downspout.

Splice. Joining of two similar members in a straight line.

Split Jamb. A door frame fabricated in two interlocking halves, to be installed from the opposite sides of an opening.

Split Ring. A connector for timbers consisting of a metal ring set in circular grooves in two pieces. The assembly is held together by bolts.

Spotlight. A light or lamp which directs a narrow intense beam of light on a small area.

Spray Booth. A room or enclosed space that provides a ventilated, dust-free environment for the application of paints.

Sprinkler Head. The outlets from which water is sprayed from a sprinkler or irrigation system.

Square. A tool used by tradesmen to obtain accuracy; a term applied to a surface which measures 10 feet by 10 feet or 100 square feet.

Stability. The ability to remain unchanged. Ability to restore to original condition after being disturbed by some force.

Staging. Temporary scaffolding with platforms placed in and around a building to create elevated work areas.

Staining. Discoloration caused by a foreign matter chemically affecting the material itself. A paint which lets the grain of wood show through.

Stainless Tubing. Material in the form of a tube constructed of stainless steel.

Stair, Access. A stair system to provide specific access to roofs, mechanical equipment rooms etc.

Stair, Concrete. A stair system constructed solely from concrete.

Stairs, Box. Stairs built between walls, and usually with no support except the wall.

Standing Finish. Term applied to the finish of the openings and the base, and all of the interior finish work.

Standing Seam Roof. A sheet metal roof system that has seams that project at right angles to the plane of the roof.

Standpipe. A pipe that extends the full height of a building, with hose connections, used to provide water exclusively for the fighting of fires. A dry standpipe does not have water in the pipe until a valve is activated. A wet standpipe is directly connected to a water supply and usually located in a stairway.

Staple. Double pointed, U-shaped metal fastener used for same purposes as nails, but providing additional head holding power.

Starter. A device that insures that a motor does not receive too high a current when starting up.

Static. State of being at rest, having no motion.

Statute of Limitations. The period of time after a cause of action arises before the expiration of which a plaintiff must file suit, or lose the right to do so.

Steam Boiler. A boiler for producing steam.

Steam Trap. A device that allows the passage of a condensate and/or air, but prevents the passage of steam.

Steel. Iron compounded with other metals to increase strength and wearing or rust resistance.

Steel Angle. An L-shaped member constructed of steel, often used as a lintel or carrying shelf for masonry.

Steel Joist. Open web, parallel chord, load-carrying members suitable for direct support of floors and roof decks, utilizing hot rolled or cold formed steel.

Steel Ladder Cage. Open steel framework enclosing ladder on open side to prevent falls from ladder. Commonly required by safety codes on high ladders.

Steel Pile. A structure element made of steel driven or embedded in the ground for the purpose of supporting a load. A long, slender, piece of steel driven into the ground to act as a foundation. A member embedded into the ground that supports vertical loads.

Steel Pin and Roller. Joint used in steel bridge construction to allow for rotation and movement.

Steel Pipe. Hollow structural tubular pipe made of steel.

Steel Pipe Fitting. Elbows, crosses, tees, caps, flanges made of steel; slip type, threaded, or flush welded to pipe to facilitate transitions and terminations.

Steel Plate. Sheet steel of a heavier thickness.

Steel Square. A tool which helps create a right angle between two components. The large arm of the square is called the body or blade. The smaller arm is at a 90-degree angle to the blade is called the tongue. The point where the outside edges of the blade and tongue join is called the heel. The surface with the manufacturer's name is called the face; the opposite surface is called the back.

Steel, Structural. Steel that is rolled in a variety of shapes and manufactured for use as structural load-bearing members.

Steel Stud. Steel pin or rod having head at one end for driving into material used for holding members or parts of members together. In light gage construction, a vertical framing member such as a "2 by 4".

Steel Sump Pan. Sheet metal pan forming low point in roof deck to collect water and receive roof drain.

Steel Tank. A receptacle manufactured from steel for holding, transporting or storing liquids.

Stick, Hook. An implement that is curved or bent for holding, catching or pulling.

Stiffener Plate. A steel plate attached to a structural member to support it against heavy localized loadings or stresses.

Stiffness. The quality of resistance to deformation on the part of a material, a component member of a structure, or the whole structure.

Stile. A vertical framing member in a panel door.

Still Water. A part of a stream where the gradient is so gentle that no current is visible.

Stirrup. A vertical loop of steel bar used to reinforce a concrete beam against diagonal tension forces.

Stone. Earthy or mineral matter of indeterminate size or shape such as rock, etc.

Stone Paver. Blocks of rock processed by shaping, cutting or sizing, used for driveways, patios and walkways.

Stoned. Use of a carborundum stone to eliminate jagged and flaked edges of tile or masonry due to cutting.

Stop, Gravel. A metal flange or strip with a vertical lip placed around the edge of a built-up roof to prevent loose gravel from falling off the roof.

Stop Notice. A charge against construction funds in the hands of a property owner or a construction lender for the value of work or materials incorporated into a construction project.

Storefront. The facade which is constructed on the street side of a building or structure into which persons can enter and transact business. A loosely used term to denote a steel or aluminum tube frame and glass wall.

Straightedge. A straight piece of lumber or metal that is used to rod mortar, align tile, or provide a straight or flush surface..

Strap Tie. A metal plate that fastens two parts together as a post, rod or beam.

Strength of Materials. A branch of mechanics and experimental physics dealing with stresses, strains, and the general behavior or materials and structural elements under the action of forces and moments.

Stress. The internal force of a body that resists external force.

Stressed-Skin Panel. A panel consisting of two face sheets of wood or metal bonded to perpendicular spacer strips.

Stretcher. A masonry unit laid with its length horizontal and parallel with the face of a wall or other masonry member. A brick or masonry unit laid in its most usual position, with the broadest surface of the unit horizontal and the length of the unit parallel to the surface of the wall.

Striking Joints. A process of removing excess grout from the joints by wiping with a sponge or cloth or scraping, compacting or rubbing with a curved instrument.

Stringer. The sloping wood or steel member supporting the treads of a stair. A long horizontal timber in a structure supporting a floor.

Strip, Cant. A strip of material, usually treated wood or fiber, with a sloping face used to ease the transition from a horizontal to a vertical surface at the edge of a flat roof.

Strip, Chamfer. An insert that is triangular or curved, placed in an inside corner to produce a rounded or flat beveled edge at the right angle corner of a construction member.

Structural Bond. The interlocking pattern of masonry units used to tie two or more wythes together in a wall.

Structural Lumber. Wood members of a structural system which are manufactured by sawing, resawing, passing lengthwise through standard planing machine, crosscutting to length, but without further manufacturing.

Structural Pipe. Pipe used in a structure to transfer imposed loads to the ground.

Structural Plywood. The highest grade of plywood. Plywood of exterior structural grade, secured to top side of floor joists used to create rigidity in building superstructure and also to provide smooth and even surface to receive finish floor covering.

Structural Steel. Steel hot-rolled into variety of shapes for use as load-bearing structural members.

Structural Tube. A hollow metal product used to carry imposed loads in a building or structure.

Strut. A compression member, a column, usually implying that it can be placed at any angle, not just vertically.

Stucco. A cement plaster used for coating exterior walls and other exterior surfaces of buildings. Portland cement plaster used as an exterior cladding or siding material. A plaster used for interior decoration and finish work; also for rough outside wall coverings.

Stucco Lath. Wood or metal lath strips which form the base for the application of a cement plaster on an exterior wall surface.

Stud. One of a set of small vertical elements, usually wood, used to produce a framed wall. An upright beam in the framework of a building. Vertical member of appropriate size (2x4 to 4x10 in.) (50x100 to 100x250 mm) and spacing (16 to 30in.) (400 to 750 mm) to support sheathing of concrete forms; also a headed steel device used to anchor steel plates or shapes to concrete members.

Stud, Metal. Vertical, formed channel of "C" steel component within a framed wall, may be either load-bearing or non-load-bearing, to withstand structural loads imposed by wind and suction loads, building loads, movement and deflection of structure.

Sub-Flooring. Certain material, like plywood, that is installed on the floor joists of a building or structure, onto which the walls and finished flooring is attached.

Subdivider. One who buys land wholesale and sells it retail.

Subfloor. Typically a wood floor which is laid over the floor joists and on which the finished floor is laid.

Subpurlin. A small roof framing member that spans between joists and purlins.

Summary Judgment. A judgment awarded on the basis of affidavits and legal briefing rather than on the basis of evidence introduced at a trial.

Sump Pump. A small capacity pump that empties pits receiving groundwater, sewage, or liquid waste.

Superintendent. An individual who is at the top level of a construction team in the field. A person with executive oversight, often of a board or building complex.

Supplier. One who supplies construction materials to a project.

Surcharge. An increase in the lateral earth pressure of a retaining wall, caused by a vertical load behind the wall. A load placed over an area to compact it or change its characteristics.

Surety. One who undertakes to guarantee performance by another. An insurance company.

Surface Waters. Rain water collected and running on the surface of the land rather than being confined to drains and water courses.

Suspended Ceiling. A finish ceiling that is hung on wires from the structure above.

Suspended Ceiling Removal. The act or process of removing the modular units and skeleton frame of an old suspended ceiling.

Suspended Structure. A structure supported principally by tension members or carrying its loads principally in tension.

Swale Excavation. The digging up of low-lying land such as a small meadow, swamp, or marshy depression.

Swing Gate. The operable member of a fence system that is hinged for opening and closing.

Switch, Fusible. An electric switch that has a fusible link in the wiring for an electric circuit in the event of an overload.

Switchboard. An apparatus consisting of a panel on which measuring, controlling, etc., devices are arranged to that a number of circuits may be measured, controlled, etc.

Switchgear. A freestanding assembly including primary (disconnect) switches, secondary (feeder) switches, and overcurrent protection device (fuses and circuit breakers).

System. An ordered assemblage.

T

T&G Siding. Tongue and groove exterior siding.

T-Beam. A reinforced concrete beam that contains of a portion of the slab above and which the two act together.

Tack. Short sharp pointed nail with large head used to secure thin or woven materials to wood and similar materials.

Tack Coat. Application of material made from asphalt on old surface to insure its bond to new construction.

Take-Off Man. Someone who can read blueprints and is familiar with the specifications. This person makes notes of special details concerning the project after gathering the necessary information and then estimates the quantities of labor, materials, equipment and special items needed to complete the job. Also may be called a Quantity Surveyor.

Tank, Toilet. The reservoir located in the back of a water closet, which holds the water necessary for flushing away waste.

Tapping Valve. A device to open or close a duct, pipe, or other passage, or to regulate flow. It is inserted into an existing pipeline by piercing the wall of the pipe and thus tapping into the flow.

Tarpaulin. A waterproofed canvas or other material used for protecting construction projects, athletic fields, goods or other exposed objects or areas.

Teak Veneer. Thin sheets of teak, a dark wood, used for plywood or other finishes.

Tee. A metal or precast concrete member with a cross section resembling the letter "T". A pipe fitting which joins three pipes at 90 degree angles.

Tee, Bulb. Rolled steel in the form of a "T" with a formed bulb at the end of the web.

Tee Weld. Weld in joint between two members located approximately at right angles to each other in form of "T".

Tempered Glass. Glass that has been treated to increase it toughness and its resistance to breakage.

Temporary Centering. Support used during construction process (not permanent) for support of masonry arch or a concrete slab.

Temporary Facility. A structure erected for temporary use.

Tendon. A steel strand used for prestressing a concrete member.

Tensile Strength. The pulling force necessary to break a given specimen divided by the cross sectional area. Units given in lbs./in.^{2} (P.S.I.). It measures the resistance of a material to stretching without rupture. Normally is not used with reference to elastic materials which recover after elongation. The ability of a structural material to withstand tensile forces.

Tension. Stress which tends to elongate a member.

Tension Ring. A structural element, forming a closed curve in plan, which is in tension because of the action of the rest of the structure. A concrete or masonry dome commonly has a tension ring.

Terne. An alloy of lead and tin, used to coat sheets of carbon steel, stainless steel, or copper for use as metal roofing sheet.

Terra Cotta. Hard baked clayware, or tile, of variable color, averaging reddish red-yellow in hue and of high saturation.

Terrazzo. A finish floor material consisting of concrete with an aggregate of marble chips selected for size and color, which is ground and polished smooth after curing.

Test. A trial, examination, observation, or evaluation used as a means of measuring a physical or chemical characteristics of a material, or a physical characteristic of a structural element or a structure.

Test Pile. A pile which is driven before the final design is done to see what bearing is developed and what length of pile is actually needed. A pile which is tested by placing a predetermined load on it, commonly done by erecting a crib on the pile and then filling it with ballast.

Testimony. Oral evidence offered by a witness in the course of a hearing, a trial, or a deposition.

Testing Machine. A device for applying test conditions and accurately measuring results.

Thermodynamics. The science dealing with the relationship between heat and other forms of energy.

Thermoplastic. Having the property of softening when heated and rehardening when cooled.

Thermostat. A device which measures temperature (typically of a space) and activates an HVAC system to heat or cool the air.

Thin Coat. A loosely used term for a one coat plaster system over gypsum board.

Thin Set. A term used to describe the bonding of tile with suitable materials, applied approximately 1/8" thick.

Threshold. The wood or metal beveled floor piece at door openings which commonly separates non-continuous floor types.

Threshold, Door. A beveled piece of floor trim over which a door swings.

Thrust. A lateral or inclined force resulting from the structural action of an arch, vault, dome, suspension structure, or rigid frame.

Tie. A device for holding components together, a structural device that acts in tension.

Tie Rod. A steel rod that acts in tension and commonly holds together wall forms while concrete is being poured.

Tile. A ceramic surfacing unit, usually relatively thin in relation to facial area, made from clay or a mixture of clay and other ceramic materials, called the body of the tile, having either a glazed or unglazed face and fired above red heat in the course of manufacture to a temperature sufficiently high to produce specific physical properties and characteristics. A fired clay product that is thin in cross section as compared to a brick; either a thin/flat element (ceramic tile or quarry tile), a thin/curved element (roofing tile), a hollow element with thin walls (masonry tile), or a pipe-like shape (drainage tile).

Tile, Acoustical. Finished ceiling panels in board form used for its sound absorbing properties. Sometimes used on walls.

Tile, Quarry. Clay tile that is fired and used for flooring.

Tile Resilient. Tile manufactured from rubber, vinyl, or other resilient materials.

Tilt-Up Construction. A method of constructing concrete walls in which panels are cast and cured flat on the floor slab, then tilted up into their final positions.

Timber. Lumber with cross-section over 4 by 6 inches, such as posts, sills, and girders.

Timekeeper. A clerk who keeps records of the time worked by employees.

Tinted Glass. Glass that has been treated to reduce transmitted glare.

Title Insurance. A guarantee of title issued by an insurance company.

Third Party Beneficiary. A person who is not a party to a contract but who is the intended beneficiary thereof and may therefore enforce it.

TLVs. Threshold Limit Values.

To the Weather. A term applied to any part of the structure which faces the elements. A shingled roof is "to the weather" the framing system is not.

Toggle Bolt. A bolt and nut assembly to fasten objects to hollow construction assembly from only one side. Nut has pivoted wings that close against spring when nut end of assembly is pushed through hole and spring open on other side in void of construction assembly.

Toilet Partition. Privacy panels in a toilet enclosure.

Tone. A sound of only a single frequency.

Tongue and Groove. A type of lumber, metal or precast concrete having matching or mated edges to provide a tight fit, abbreviated "T & G".

Toothed Ring. A timber connector used in the manufacturing of large member wood trusses.

Top Bars. Steel reinforcing bars near the top of reinforced concrete.

Top Plate. The horizontal member at the top of a stud wall, usually supporting rafters.

Topping, Concrete. Concrete layer placed to form a floor surface on a concrete base.

Topsoil. Surface soil at and including the average plow depth, soil which is used as a planting or growing medium.

Torque. Twisting action; moment.

Torsion. The rotation of a diaphragm caused by lateral forces and whose center of mass does not coincide with the center of rigidity.

Tort. A negligent or intentional wrongful act that damages the person or property of another, the wrongful nature of which is independent of any contractual relationship.

Total Cost. A method of computing damages sustained by a contractor because of breaches of contract causing the contractor to operate in an inefficient or unproductive manner.

Towel Dispenser. A container that holds paper towels for future use, and which has an opening allowing one towel to be removed at a time.

Tower. A tall structure, constructed of frames, braces, and accessories rising to a greater height than the surrounding area.

Trailer. A vehicle designed to be hauled or to serve parked as dwelling or place of business.

Trammel Bar. A tile layout tool. It is used to erect perpendicular lines and to bisect angles.

Transformer. A device that changes, or transforms, alternating current from one voltage to another.

Transformer Pad. A precast concrete block or stone placed under a transformer to spread and support its weight.

Transit Mixed Concrete. Concrete mixed in a drum on the back of a truck as it is transported to the structure.

Translate. To move from one place to another, without rotation. Motion of a body along a line without rotation or turning.

Translucent Panel. A building panel that permits the passage of light but not vision.

Transom. A transverse piece in a structure, lintel. A horizontal crossbar in a window, over a door, or between a door and a window or a fanlight above it. A window above a door or other window built on and commonly hinged to a transom.

Transverse. Across, at right angles to the main axis of a structural member.

Trap. Located at a plumbing fixture, designed to hold a quantity of water that prevents gasses in the sewer system from entering a room.

Trash Chute. A device either constructed on the inside of a building or structure or hung outside for the removal of waste materials from upper floors.

Traveling Crane. A tower crane mounted on tires, rails or crawlers.

Tray, Cable. Open track support for insulated cables.

Tread. One of the horizontal planes of a stair. The horizontal part of a step.

Treated Lumber. Lumber infused or coated with stain or chemicals to retard fire, decay, insect damage or deterioration due to weather.

Tremie. A large funnel with a tube attached, used to deliver concrete into deep forms or beneath water or slurry. A tremie slows down the concrete and resists segregation of the aggregates.

Trench. A long, thin excavation.

Trencher. A mechanical device used to dig narrow channels in the ground.

Trenching. The act or process of digging narrow channels in the ground.

Trespass. Unauthorized entry upon the real property of another.

Trial Batch. A batch of concrete prepared to establish or check proportions of the constituents.

Trim. A term sometimes applied to outside or interior finish woodwork and the finish around openings.

Trimmer. The beam or floor joist into which a header is framed.

Troffer. A channel-like enclosure for light sources.

Trowel Finish. The final finish of concrete, plaster, stucco, etc., by the use of a hand trowel.

Truck. A wheeled vehicle for moving heavy articles. A strong automotive vehicle for hauling. A small wheelbarrow consisting of a rectangular frame having at one end a pair of handles and at the other end a pair of small, heavy wheels and a projecting edge to slide under a load.

Truck Crane. A mechanical device for hoisting or lifting materials mounted on the bed of a truck.

Truss. A triangular arrangement of structural members that reduces nonaxial forces on the truss to a set of axial forces in the members. Structural framework of triangular units for supporting loads over long spans.

TSI. Thermal System Insulation, insulative material applied to pipes, fittings, boilers, tanks, ducts or other interior structural components to prevent heat loss or gain or water condensation.

Tube, Structural. A hollow metal or plastic product used to carry imposed loads in a building or structure.

Tuck Pointing. The process of removing deteriorated mortar from the surface of an existing brick wall, and inserting fresh mortar.

Tunneling. The act or process of digging a horizontal passageway through or under an obstruction.

Turnout. A widened space in a highway for vehicles to pass or park. A railroad siding.

TVOC. Total volatile organic compound.

TWA. Time-weighted Average. In air sampling, this refers to the average air concentration of contaminants during a particular sampling period.

Two-Way Flat Slab. A reinforced concrete framing system in which columns with mushroom capitals and/or drop panels directly support a two-way slab that is planar on both its surfaces.

Type X Gypsum Board. A gypsum board used where increased fire resistance is required.

U

U Bolt. A U-shaped, bent iron bar that has bolts and threads at both ends.

U Stirrup. An open-top, U-shaped loop of steel bar used as reinforcing against diagonal tension in a beam.

UHF Cable. Cable that is designed for ultra-high frequency.

Ultimate Load. The absolute maximum magnitude of load which a structure can sustain, limited only by ultimate failure.

Ultimate Strength. Maximum strength that can be developed in a material.

Ultimate-Strength Design. Method of structural analysis of continuous concrete structures based on calculating the loadings which will cause actual failure of sections, rather than the loadings which will cause stresses to reach allowable, safe, values.

Ultrasonic Transmitter. A mechanism that converts vibrations with the same physical nature as sound into equivalent waves above the range of human hearing.

Unbonded Construction. Post-tensioned concrete construction in which the tendons are not grouted to the surrounding concrete.

Uncertainty. The doctrine that contracts are unenforceable if unintelligible

Unconscionability. A doctrine that courts will not enforce contractual provisions that put one party at the mercy of another.

Undercourse. A course of shingles laid beneath an exposed course of shingles at the lower edge of a wall or roof, in order to provide a waterproof layer behind the joints in the exposed course.

Underdrain. Perforated pipe drain installed in crushed stone under a slab to intercept ground water and drain it away from the structure.

Underfloor Duct. A round or rectangular metal pipe placed in a concrete floor to distribute warm air from a heating or air conditioning system.

Underground Tank. A container that holds various liquid or solid matter and is found underground.

Underlayment. A layer of hardboard, particleboard, plywood, mastic with latex binders, mastic with asphalt binders, mastic with polyvinyl-acetate, etc., placed to cover subfloor irregularities, to absorb the movement of wood subfloors and to provide a smooth surface for the finish flooring material.

Underpinning. The process of placing new foundations beneath an existing structure.

Underslab Drainage. The process of continuous interception and removal of ground water from under a concrete slab with the installation of perforated pipe.

Ungrounded Cable. Two-wire nonmetallic sheathed cable that contains one neutral wire and one hot wire.

Uniform Commercial Code. A statute governing commercial transactions, sales, and commercial paper that has been adopted in substantially the same form in all 50 states.

Union. A type of pipe fitting used to join two pipes in line without turning either pipe.

Union T. A pipe tee with a fitting on one end that joins two pipes without turning either pipe.

Unit Heater. A device for heating a space without the use of ductwork.

Unit Masonry. Manufactured or natural building units of concrete, burned clay, glass, stone, gypsum, etc..

Unit Substation. A freestanding assembly including a transformer, switchgear and meter(s).

Unit Ventilator. A fan coil unit (FCU) with an opening through an exterior wall to admit outside air.

Unit-and-Mullion System. A curtain wall system consisting of prefabricated panel units secured with site-applied mullions.

Unitary Package Unit. A system of conditioning air in an outdoor package unit, similar to a rooftop unit, except that the supply and return ductwork passes horizontally through an outside wall.

Unjust Enrichment. A situation in which one party is unjustly enriched at the expense of another party.

Unlined Stack. Chimney or vent fabricated of single pipe.

Unrated Door. A door which has not been rated for any ability to withstand the spread of fire.

Unreinforced. Constructed without steel reinforcing bars or welded wire fabric.

Uplift. An upward force.

Upside Down Roof. A membrane roof assembly in which the thermal insulation lies above the membrane.

Urethane Board. Rigid form of plastic foam of polyurethane.

Urinal. A plumbing fixture used to collect urine and which can be flushed.

Urinal Screen. A privacy panel that separates urinals from each other. The screens are either floor, wall or ceiling mounted.

Usury. Collecting, or contracting to collect, interest at higher than the lawful rate.

Utility. A service provided by a public utility (as light, power, or water).

Utility Excavation. The act or process of either digging up existing cable buried in the ground, or trenching to lay new cable.

Utility Pole. A vertical pole with cross arms, where utility company lines are carried.

V

V. A symbol for vertical shear.

V Beam Roof. A roof with corrugated sheeting with flat, V-angled surfaces.

Vacuum Breaker. An electrical breaker with a space that contains reduced air pressure.

Valley. A trough or internal angle formed by the intersection of two roof slopes. The internal angle formed by the two slopes of a roof.

Valley Flashing. Thin sheet metal used to line the valley of a roof.

Valley Rafter. A diagonal rafter that supports a roof valley.

Valve. Numerous mechanical devices by which the flow of liquid, gas, or loose material in bulk may be started, stopped, or regulated by a movable part that opens, shuts, or partially obstructs one or more passageways.

Vanity Cabinet. Case, box, or piece of furniture which sets on floor and receives a lavatory, commonly has shelves and doors and is primarily used as storage for below lavatory.

Vapor Barrier. Waterproof membrane placed under concrete floor slabs that are on grade.

Varnish. A colorless, clear, resinous product dissolved in oil, alcohol, or a number of volatile liquids, applied to wood to provide a hard, glossy, protective film.

Vault. An arched surface. An arch translated along an axis normal to the plane of its centerline curve. A room to store valuable items.

VAV. Variable air volume system.

Veneer. A thin layer, sheet or facing.

Veneer, Ashlar. An ornamental or protective facing of masonry composed of squared stones.

Veneer Plaster. A wall finish system in which a thin layer of plaster is applied over a special gypsum board base.

Vent. A vertical pipe connected to a waste or soil distribution system that prevents a vacuum that might suck the water out of a trap. Vertical pipe to provide passageway for expulsion of vent gases from gas-burning equipment to outside air.

Vent, Foundation. Opening in foundation wall to provide natural ventilation to foundation crawl spaces.

Vent Stack. A plumbing vent (pipe) in a multistory building, a separate pipe used for venting, that either connects with a stack vent above the highest fixture, or extends through the roof.

Ventilation Air. Defined as the total air, which is a combination of the air brought into the system from the outdoors and the air that is being recirculated within the building. Sometimes, however, used in reference only to the air brought into the system from the outdoors.

Ventilator, Gravity. A device installed in an opening in a room or building which is activated by air passing through to remove stale air and replace it with fresh air.

Verge Boards. The boards which serve as the eaves finish on the gable end of a building.

Vermiculite. Expanded mica, used as an insulating fill or a lightweight aggregate.

Vermont Slate. A fine grained thin-layered rock used for roofing, paving, etc..

Vertical Bar. An upright reinforcing bar in a reinforced concrete shape.

Vertical Siding. Exterior wall covering attached vertically to the wood frame of a building or structure.

Vestibule. An entrance to a house; usually enclosed.

Vibration. A periodic motion which repeats itself after a definite interval of time.

Vierendeel "Truss" (or Frame). A rigid frame, used as a beam, assembled from parallel top and bottom chords tied together by vertical members.

Vinyl Sheet. The rolled form of vinyl.

Vinyl Sheetrock. Plasterboard with a thin layer of vinyl as the finished surface.

Vinyl Siding. Exterior wall coverings made from a thermoplastic compound.

Vinyl Tile. A semi-flexible, resilient floor tile made from polymerized vinyl chloride, vinylide chloride, or vinyl acetate.

Vision Panel. Glass placed in an opening of a door.

Vitreous (Vitrified). That degree of vitrification evidenced by low water absorption. The term vitreous generally signifies less than 0.5% absorption, except for floor and wall tile and low-voltage electrical porcelain which are considered vitreous up to 3.0% water absorption.

Vitrified Clay Pipe. Pipes used especially for underground drainage, that are made of clay baked hard.

VOCs. Volatile Organic Compounds.

Void. An unfilled space in a material of trapped air or other gas.

Voltage. Electricity is caused by creating a higher electric charge at one point in a conductor than at another. This potential difference is called voltage.

W

Waferboard. A building panel made by bonding together large, flat flakes of wood.

Waffle Slab. A two-way concrete joist system. Two-way slab or flat slab made up of a double system of narrow ribs or joists, usually at right angles to each other, forming a pattern of waffle-like coffers.

Wainscoting. A wall facing, usually of wood, cut stone, or ceramic tile, that is carried only part way up a wall. Matched boarding or panel work covering the lower portion of a wall.

Waiver. The intentional relinquishment of a known right.

Wale. A horizontal beam.

Walk. A path specially arranged or paved for walking.

Walkway, Roof. A permanent aisle with handrails for safe access across a roof.

Wall. A member, usually vertical, used to enclose or separate spaces.

Wall Blocking. Framing lumber cut in short lengths and installed horizontally between wall studs as filler pieces to stabilize the framing or to provide a backing for fastening a finish item.

Wall Flange. A ridge on a wall that prevents movement. A supporting rim on a wall for attachments.

Wall Footing. A continuous spread footing that supports a uniform load from a wall.

Wall Heater. A heating unit installed in a wall.

Wall Hydrant. A connection to a water main cut through and mounted on a wall.

Wall, Retaining. A wall that is designed to resist the lateral pressures of retained soil.

Wall Sheathing. The first layer of covering on an exterior wall, fastened to the wall studs.

Wall Tie. A mechanical metal fastener which connects wythes of masonry to each other or to other materials.

Walnut Veneer. An overlay of a thin layer of walnut wood for outer finish or decoration.

Wardrobe. A room or freestanding closet where clothes are kept.

Wardrobe Locker. A cabinet with a locking door where clothes are kept.

Warpage. A concave or convex curvature of a material that was intended to be perfectly flat.

Wash. The slant upon a sill, capping, etc., to allow the water to run off easily.

Wash Fountain. A waist high sink which supplies a steady stream of water to cleanse the hands.

Washer. A flat thin ring or a perforated plate used in joints or assemblies to ensure tightness or relieve friction.

Washer, Flat. A washer which goes under a bolt head or a nut to spread the load and protect the surface.

Washroom Faucet. A device that dispenses hot and cold water, mounted above a sink.

Waste Stack. A vertical pipe that carries liquids other than human waste.

Watchman. A guard who keeps watch over a certain area.

Water Absorption. The ability to take up and retain water.

Water Closet. A plumbing fixture for the disposal of human wastes, a toilet.

Water Cooler. An apparatus that holds and dispenses cold water.

Water Hammer. Noise, occurring in a water pipe, when air is trapped in the pipe.

Water Heater. An apparatus for storing and heating water.

Water Level. A basic device to check level in walls or structures. Commonly, a length of clear plastic hose 3/8" to 1/2" in diameter and approximately 50 feet long. It is filled with water and used as a leveling device.

Water Meter. A device for measuring the flow of water.

Water Reducing Admixture. Material added to cement or a concrete mix to cut down on its water content.

Water Slide. A sloping trough down which water is carried by gravity.

Water Softener. A device attached to a water system to remove unwanted minerals and substances.

Water Table. The level at which the pressure of water in the soil is equal to atmospheric pressure; effectively, the level to which ground water will fill an excavation. The finish at the bottom of a house which carries water away from the foundation. A projection on the bottom of an exterior wall to prevent rain or water from seeping through to the wall below.

Water Valve. A device to regulate the flow of water in a pipe or other passage.

Water Cement Ratio. The ratio of the amount of water, exclusive only of that absorbed by the aggregates, to the amount of cement in a concrete or mortar mixture; preferably stated as a decimal by weight. A numerical index of the relative proportions of water and cement in a concrete mixture.

Water Resistant Gypsum Board. A gypsum board designed for use in locations where it may be exposed to occasional dampness. Plasterboard that has had a chemical treatment to make it resistant to moisture, but not necessarily waterproof.

Waterproof Membrane. A membrane, which can be made of built-up roofing or an elastomeric sheet, to provide positive waterproofing for a floor wall.

Waterproofing. The act or process of making something waterproof. A coating capable of stopping penetration of water or moisture.

Waterstop. A synthetic rubber strip used to seal joints in concrete foundation walls.

Watertight Manhole. A cover for a vertical access shaft that prevents the elements from coming in.

Wave. Any disturbance that advances through a medium with a speed that is completely determined by properties of that medium, i.e., sound, light.

Wave Length. The distance between successive similar points on two wave cycles.

Wax, Floor. A substance spread onto flooring that seals, protects and can be polished to a shine.

Wearing Course. A topping or surface treatment to increase the resistance of a pavement or slab to abrasion.

Weather Seal. A flanged channel installed on the edges of an exterior door.

Weathered Joint. A mortar joint finished in a sloping profile that tends to shed water to the outside of the wall.

Weathering. Changes in color, texture, strength, chemical composition or other properties of a natural or artificial material due to the action of the weather.

Weathering Steel. A steel alloy that forms a tenacious, self-protecting rust layer when exposed to the atmosphere.

Weatherproof Box. An electrical outlet or switch box which has been manufactured to withstand the outside elements.

Weatherstrip. A ribbon of resilient material used to reduce air infiltration through the crack around a sash or door.

Weatherstripping. The process of reducing air or rain infiltration by covering joints of doors or windows with strips of resilient material.

Web. The vertical plate connecting the top and bottom flanges of a beam.

Web (Masonry). An interior solid portion of a hollow concrete block (CMU).

Weed Control. The act or process of spraying chemicals or placing powders to control the spread of weeds.

Weep Hole. A small opening, the purpose of which is to permit drainage of water that accumulates inside a building component.

Weld. A joint between two pieces of metal formed by fusing the pieces together, usually with the aid of additional metal melted from a rod or electrode. Join two pieces of metal together by heating until fusion of material either with or without filler metal.

Weld, Destructive Test. Methods to determine existence and extent of defects and discontinuities in welds which do affect capabilities of weld and require repairs after testing.

Weld Inspection. Methods to determine existence and extent of defects and discontinuities in welds.

Weld Plate. A steel plate anchored into the surface of concrete, to which another steel element can be welded.

Weld Test. Generally used as term synonymous with Welding Examination (Inspection). The loading of welds to determine load capacity (similar to testing compressive strength of concrete cylinders) of welds is not normal practice.

Weld X-Ray. To examine, treat, or photograph the connection of surfaces that have been welded together.

Welded Pipe. Piping where connections and fittings are welded.

Welded Railing. Railing sections with the components fastened with welds.

Welded Truss. Trusses with components fastened together with welds.

Welded Wire Fabric. A series of longitudinal and transverse wires arranged substantially at right angles to each other in sheets or rolls, used to reinforce mortar and concrete.

Welded Wire Fabric Reinforcement. Welded-wire fabric in either sheets or rolls, used to reinforce mortar and concrete.

Welder. One who is capable of performing manual or semiautomatic welding operations based on training, experience, testing, or certification, or any combination of these.

Welding. Fusing metallic parts by heating and allowing the metals to flow together.

Welding Electrodes and Rod. The electrode and rod are the components of the welding circuit through which current is conducted between the electrode holder and the arc.

Welding Inspector. One who is capable of inspection of welds based on training, experience, testing, or certification, or any combination of these.

Welding Test. The act or process of testing the strength of a weld.

Weldment Connection. The assembling together of pieces by welding to create a unit.

Well. A pit or hole sunk into the earth to reach a supply of water. An open space extending vertically through floors of a structure.

Well Graded Aggregate. Aggregate having a particle size distribution which will produce maximum density, i.e., minimum void space.

Wellpoint. A perforated pipe surrounded by sand to permit the pumping of ground water.

Wellpoint System. A series of vertical pipes in the ground connected to a header and pump to drain marshy areas or to control ground seepage.

Wet Areas. Interior or exterior tiled areas subject to periodic or constant wetting. Examples: showers; sunken tubs; pools; exterior walls; roofs; exterior paving and interior floors.

Wet Location Fluorescent. A watertight fluorescent fixture that is sealed to protect against moisture.

Wet Sprinkler System. A sprinkler system that is filled with water at design pressure for immediate use upon activation.

Wetting. The thorough impregnation of a material by a liquid. The more viscous a fluid, and the higher its surface tension, the more difficult it is for the liquid to "wet" materials. Certain additives, for example, water softeners, reduce surface tension, or viscosity and improve wetting properties, allowing the material to better flow.

Wetting Agent. A substance capable of lowering the surface tension of liquids, facilitating the wetting of solid surfaces and permitting the penetration of liquids into capillaries.

Wharf. A structure that provides berthing space for vessels, to facilitate loading and discharge of cargo.

Wheel Barrow. A small vehicle with handles and one or more wheels used for carrying small loads.

Wheel Chair Partition. A dividing wall in a bathroom or bathing room which forms the perimeter of a private area that has been made accessible to the disabled.

Wheeled Extinguisher. A fire extinguisher mounted on a wheeled cart that can be pushed or pulled by a person.

White Cedar Shingle. A light colored weather-resistant cedar wood used for roofing and siding.

White Cement. Cement made from materials with low iron content to produce mortar or concrete that is white in color.

Whiting. Calcium carbonate powder of high purity.

Wide Stile Door. Wider than normal vertical members forming the outside framework of a door.

Wide Flange Section. Any of a wide range of steel sections rolled in the shape of a letter "T" or "H".

Wind ("i" pronounced as in "kind"). A term used to describe the surface of a board when twisted (winding) or when resting upon two diagonally opposite corners, if laid upon a perfectly flat surface.

Wind Brace. A diagonal structural member whose function is to stabilize a frame against lateral forces.

Wind Drift. Horizontal deflection of a frame caused by wind forces.

Wind Load. A load on a building caused by wind pressure.

Wind Uplift. Upward forces on a structure caused by negative aerodynamic pressures that result from certain wind conditions.

Wind Restraint System. The collection of structural elements which provide restraint of the seismic-isolated structure for wind loads. The wind-restraint system may be either an integral part of isolator units or may be a separate device.

Winder. A stair tread that is wider at one end than at the other.

Window. An opening in the wall of a building or structure for the admission of light and air, closed by casements or sashes containing glass panes.

Window, Drive-Up. An opening in a wall through which transactions can be made with persons in motor vehicles.

Window Frame. The structure which holds a window assembly in place.

Window Guard Lock. Tamperproof hasp and padlock for window guards.

Window Guard, Diamond Mesh. Guard fabricated of diamond-shaped mesh to provide protection over the face of a window to prevent damage to glass and/or to prevent intrusion.

Window Guard, Steel Bar Grille. Guard fabricated from a steel bar grille.

Window Guard, Woven Wire. Guard fabricated of woven wire..

Window Header. A horizontal construction member placed across the top of a window opening to support the load above.

Window, Service. An opening in a wall or partition through which business is transacted.

Window Sill. The horizontal member at the bottom of a window.

Window Sill, Marble. Marble installed on the horizontal member at the bottom of a window.

Window, Steel. An assembly installed in an opening in an external wall consisting of glass and a steel frame.

Window Stool. Wood, ceramic tile, or masonry installed on the plate at a window sill on the inside of the window, fitted against the bottom rail of the lower sash.

Window Sub-Sill. Component anchored to wall construction and located just below window sill to receive window sill.

Window Treatment. The addition of hanging fabrics, curtains, or blinds to the interior of a window.

Window Well. Recess located at or below grade to allow for natural light to reach a ground level or basement window, often created by use of corrugated metal in half-round shape.

Windowwall. The opening in a wall surface which contains a window assembly or wall of assemblies. Often referred to loosely as tube-framing or storefront.

Wire, Aluminum. Electrical conductors and cable manufactured from aluminum.

Wire, Chicken. Thin, woven wire mounted on an exterior wall as a base for stucco plaster.

Wire Glass. Glass in which a wire mesh was embedded during manufacture.

Wire Guard. Flexible strands of metal that have been manufactured into a unit to act as an enclosure around moving parts of machinery, around an excavation, equipment, or materials to prevent injury to the operator.

Wire Hanger. A wire that supports or connects material.

Wire Hook Hanger. Flexible strand of metal in the shape of a hook to hold a construction member in place.

Wire Mesh. A series of longitudinal and transverse wires arranged substantially at right angles to each other sheets or rolls, used to reinforce mortar and concrete.

Wire Mesh Partition. Dividing wall constructed of metal framing and wire mesh.

Wire, Stranded. Fine wires twisted together in a group to create a larger stronger cable or wire.

Wireway. A sheet metal trough with hinge or removable cover to carry several electrical cables.

Wiring, Flexible. Electrical wiring that permits ease of installation and movement from expansion, contraction, vibration, and/or rotation.

Witness. A person who observed events and is called to testify concerning those events at a hearing.

Wobble Friction. In prestressed concrete, friction caused by unintended deviation of prestressing sheath or duct from its specified profile.

Wood Anchor. A bolt or fastening device which attaches wood to wood or wood to another material.

Wood and Plastics. A category of the CSI Masterformat which is represented in Division 6 of the format. Commonly called just "Wood" or "Carpentry".

Wood Attic Catwalk. Flat wood horizontal boards secured to top of ceiling joists or bottom chords of roof trusses to provide a walkway.

Wood Base. Wood strip applied to the base of a wall to protect wall surface and finish the intersection of wall and floor.

Wood Batten. Wood strips covering vertical joints on boards used as exterior siding.

Wood Beam. Horizontal wood structural member that supports uniform and/or concentrated loads.

Wood Blocking. Small pieces of wood used to secure, join, or reinforce members, or to fill spaces between members.

Wood Board Roof Sheathing. Wood board material placed diagonally and secured to exterior side of roof rafters or trusses used to create rigidity in building superstructure and serve as base to receive roofing.

Wood Board Subflooring. Wood board material placed diagonally and secured to top side of floor joists used to create rigidity in building superstructure and serve as base to receive flooring.

Wood Bottom Plate. A flat horizontal member, also called a mudsill, that supports the vertical wall studs and posts. Horizontal wood lumber member at bottom of wall studs which ties them together and supports studs, and which rests on the sill or joists.

Wood Bridging. Diagonal or longitudinal wood members used to keep wood joist members properly spaced, in lateral position, vertically plumb, and to distribute load.

Wood Buck. Wood frame typically built into concrete or masonry wall to accommodate finish door frame.

Wood Bumper. Wood component used to absorb impact and prevent damage to other surfaces.

Wood Cant Strip. Sloped wood strip used at perimeter of roofing membrane to transition membrane from horizontal to vertical surface.

Wood Cap. Wood member used on top of an assembly to provide termination and/or finish.

Wood Carriage. Sloping beam installed between stringers to support steps of wood stair.

Wood Casing. Wood exposed millwork or trim molding around doors, windows, beams, etc.

Wood Catwalk. Catwalk or ramp made of wood often found in attic spaces.

Wood Ceiling Joist. Horizontal framing member of a ceiling made of framing lumber.

Wood Chip Mulch. Wood chips, spread on the ground to prevent erosion, control weeds, minimize evaporation and improve the soil.

Wood Column. Vertical wood structural member, usually supporting a beam.

Wood Cornice. Horizontal wood molding that may be combination of several shaped pieces.

Wood Decking. Plywood, lumber, or glued laminated member placed over roof or floor structural members for structural rigidity of building frame and to provide a surface for traffic or substrate for roofing or flooring system.

Wood Diagonal Bracing. Diagonal wood member used to prevent bucking or rotation of wood studs.

Wood Door. Hinged wood assembly used to close an opening in a wall.

Wood Door Frame. Wood members around door opening (jambs and head) upon which door is hung and within which it fits when closed.

Wood Fascia. Flat vertical wood member of cornice, eaves, or gable or other finish, generally that part of the assembly to which the gutter is secured.

Wood Fence Removal. The act or process of knocking down fence boards, rails and posts.

Wood Fiber Insulation. Wood particles and fiber used to reduce heat transfer.

Wood Fiber Panel. Form sheathing manufactured from glued and bonded wood particles.

Wood Fiber Tile. Flat pieces of covering manufactured from glued and bonded wood particles.

Wood Finish Concrete. The act or process of using a wood float to smooth irregularities left in curing concrete, work the surface or compact the concrete.

Wood Firestop. Small pieces of wood used to fill spaces between framing members to slow the spread of fire in framing cavities.

Wood Float. The wood float is sometimes used in place of the flat trowel for floating mortar. It is good for smoothing small irregularities left on the mortar bed, working the surface of the mortar before troweling on the pure coat, or compacting floor and deck mortar.

Wood Floor Joist. Horizontal structural member of a framed floor.

Wood Floor Removal. The act or process of tearing up old wood floors.

Wood Flooring. Floor coverings consisting of dressed and finished boards.

Wood Frame. Floors, roofs, exterior and bearing walls of a building or structure constructed with wood.

Wood Furring. Strips of wood applied to surfaces (usually concrete or masonry) to provide a planar surface and to provide a fastening base for finish material.

Wood Girder. Large horizontal wood beam which supports concentrated loads at isolated points along its length such as the support of joists or rafters.

Wood Ground. Narrow wood strips used around openings and at the perimeter to provide a guide for strike off of plaster to provide a straight and level or plumb line for plaster.

Wood Handrail. A bar of wood, supported at intervals by posts, balusters, or similar members, used to provide persons with a handhold.

Wood Header. Wood member placed across joist ends or at openings in a wall to support joists or studs at openings in a framing system.

Wood Hip Rafter. Sloping wood supporting member at intersection of sloping roof planes. A structural member of a roof forming the junction of an external roof angle or, where the planes of a hip roof meet.

Wood Hook Strip. Strip wood secured to wall surface to provide mounting surface for hooks.

Wood Industrial Floor. The floor in an industrial structure, constructed of wood. The main characteristic is that the flooring is durable and able to withstand wear. Industrial flooring is commonly either decking (2 inches thick) or wood blocks laid on end in a mosaic to expose the end grain and withstand heavy use.

Wood Joist. Horizontal framing member of a floor, ceiling or flat roof.

Wood Lintel. Wood header over openings in wood frame wall construction.

Wood Molding. Wood strip with curved or projecting surface, used for decorative purposes.

Wood Nailer. Strip of wood attached to steel or concrete to facilitate making nailed connections.

Wood of Natural Resistance. The heartwood of the species set forth below. Corner sapwood is permitted on 5 percent of the pieces provided 90 percent or more of the width of each side on which it occurs is heartwood. Decay resistant - Redwood, Cedars, Black Locust. Termite resistant - Redwood, Eastern Red Cedar.

Wood Pile. A long, slender wooden pole driven into the ground to act as a foundation. A member embedded into the ground that supports vertical loads.

Wood Plate. Horizontal wood lumber member on top or bottom of wall studs which ties them together and supports studs, joists, or rafters.

Wood Pole. A long piece of wood used to carry utility lines.

Wood Rafter. Sloping wood framing member of roof immediately beneath sheathing.

Wood Rail. Wood horizontal member, often used as trim or a member supported by vertical posts.

Wood Railing Baluster. Small vertical wood member to support railing.

Wood Railing Newel. Principal wood post at foot of stairway or central support of a winding flight of stairs.

Wood Railing Post. Large vertical wood member to support railing.

Wood Ridge Rafter. Horizontal wood supporting member at top of sloping roof immediately beneath sheathing.

Wood Riser. The vertical wood board under the tread in a stairway system.

Wood Roof Cricket. Relatively small elevated area of wood roof constructed to divert water around chimney, curb, or other projection.

Wood Roof Curb. Wood member elevated above plane of roof surface used for mounting of equipment or other elements.

Wood Roof Decking. Plywood, lumber, or glued laminated member placed over roof structural members for structural rigidity of building frame and to proved surface for traffic or substrate for roofing system.

Wood Roof Edge Strip. Wood strip (usually in plane of roof insulation) at perimeter of roof secured to structural roof deck used for securing roofing membrane.

Wood Roof Nailer. Wood strip (usually in plane of roof insulation) secured to structural roof deck used for securing roofing membrane.

Wood Saddle. Short horizontal wood member set on top of wood column to serve as seat for a girder.

Wood Screw. A screw for fastening objects in wood.

Wood Seat. Seat made of wood, plywood, or particleboard.

Wood Shake. A hand-split shingle.

Wood Shelving. Horizontal mounted surface of wood, plywood, or particleboard upon which objects may be stored, supported, or displayed.

Wood Shingle. Factory cut and shaped roof covering of wood (usually of cedar), cut into modular lengths, widths, and rectangular profile.

Wood Siding. Wood used as exterior surface or cladding for exterior framed wall to provide protection from the elements.

Wood Sill Plate. Horizontal wood lumber member on bottom of wall studs which ties them together and rests on concrete or masonry. The horizontal timbers of a house which either rest upon the masonry foundations or, in the absence of such, form the foundation.

Wood Sleeper. Wood member laid on concrete floor to support and receive fastening of wood subfloor or finish flooring.

Wood Soffit. The wooden underside of a part or member in a building, such as the under surface of an arch, beam, cornice, or stairway.

Wood Stair Finish. Exposed-to-view materials of wood stair construction.

Wood Stair Framing. Wood structural members supporting stairs or stair openings.

Wood Stair Landing. Level platform between two flights of wood stairs used to break length of single flight or to change direction of stairway.

Wood Stair Railing. Wood protective bar placed at convenient distance above stair for handhold.

Wood Stair Riser. Vertical or inclined face of a wood step, extending from the back edge of one tread to the leading edge of the tread above it.

Wood Stair Stringer. Wood member upon which stair treads bear.

Wood Stair Tread. The horizontal surface of a wooden stairway system, a step.

Wood Threshold. Wood strip fastened to floor beneath door.

Wood Top Plate. Horizontal wood lumber member on top of wall studs which ties them together and supports joists or rafters.

Wood Treatment. The act or process of applying a variety of stains or chemicals to retard fire, decay, insect damage or deterioration, due to the elements.

Wood Trim. Wood millwork, primarily moldings, used to finish off and cover joints and openings.

Wood Truss. A structural component formed of wood members in a triangular arrangement, often used to support a roof.

Wood Truss Joist. Joists of rigid open framework construction with top and bottom chords, fabricated of wood web and chord members or combination wood chord members with metal web members.

Wood Truss Rafter. Truss where chord members also serve as rafters and ceiling joists.

Wood Valley Rafter. Sloping wood supporting member at intersection of sloping roof planes. A diagonal rafter that supports a valley

Wood Wall. A vertical structure member made of wood which encloses, divides, supports or protects a building or room.

Wood Window. A unit installed in an opening in a building or structure, for light and ventilation, manufactured from wood and glass.

Woodwork, Architectural. A higher than average feature of finish work using wood for ornamental design.

Workability. The property of freshly mixed concrete or mortar which determines the ease and homogeneity with which it can be mixed, placed, compacted, and finished.

Workbench. Table at which work is accomplished.

Workers Compensation. A system established by statute under which employers are responsible for medical expenses and disabilities of workers injured while on the job. Compensation is payable even if the employer is not at fault and even if the carelessness of the worker contributed to the injury, but the employer is not necessarily liable for damages for pain and suffering.

Working Drawings. Drawings of the project that are used in the construction of structure, they are part of the contract documents.

Working Stress. The maximum permissible stress used in the working stress design of a member.

Wrongful Death. Unlawful homicide, whether by negligence or intent.

Wye. A pipe fitting which joins three pipes at a 45 degree angle.

Wythe. A vertical layer of masonry one masonry unit thick.

X,Y,Z

X Bracing. That form of bracing wherein a pair of diagonal braces cross near mid-length of the bracing members.

X-Ray. Electromagnetic radiations of a short wavelength that can penetrate various thicknesses of all solids.

X-Ray, Weld. To examine, treat, or photograph the connection of surfaces that have been welded together.

Y Strainer. A device in the shape of a "Y" for withholding foreign matter from a flowing liquid or gas.

Yield Point. The point at which a material deforms with no increase in load. The stress at which a material ceases to deform in a fully elastic manner.

Z. A numerical coefficient used in the design of earthquake forces and that is dependent upon site location.

Z Tie, Wall. A Z-shaped reinforcing strip used as a support bracket from the structural wall to the masonry veneer.

Zero Slump Concrete. A concrete mixed with so little water that it has a slump of zero when tested.